REFIGURING LIFE

REFIGURING LIFE

Metaphors of Twentieth-Century Biology

EVELYN FOX KELLER

The Wellek Library Lecture Series at the University of California, Irvine

COLUMBIA UNIVERSITY PRESS NEW YORK

COLUMBIA UNIVERSITY PRESS   NEW YORK   CHICHESTER, WEST SUSSEX

COPYRIGHT © 1995 COLUMBIA UNIVERSITY PRESS   ALL RIGHTS RESERVED

The cartoon in chapter 3, note 9, is reprinted with permission from Nature 362 (April 15, 1993): 586. Copyright © 1993 Macmillan Magazines Limited.

LIBRARY OF CONGRESS CATALOGING-IN-PUBLICATION DATA

KELLER, EVELYN FOX

REFIGURING LIFE : METAPHORS OF TWENTIETH-CENTURY BIOLOGY /

EVELYN FOX KELLER.

   P.   CM.   (THE WELLEK LIBRARY LECTURES AT THE UNIVERSITY OF

CALIFORNIA, IRVINE)

INCLUDES BIBLIOGRAPHICAL REFERENCES AND INDEX.

ISBN 0-231-10204-6

I. GENETICS — HISTORY.   2. BIOLOGY — TECHNOLOGICAL INNOVATIONS.

I. TITLE.   II. SERIES.

QH428.K45   1995

575. 1'09 — DC20   94-44222

                                                          CIP

CASEBOUND EDITIONS OF

COLUMBIA UNIVERSITY PRESS BOOKS

ARE PRINTED ON PERMANENT

AND DURABLE ACID-FREE PAPER

Printed in the United States of America

C 10 9 8 7 6 5 4 3 2

EDITORIAL NOTE

The Wellek Library Lectures in Critical Theory are given annually at the University of California, Irvine, under the auspices of the Critical Theory Institute. The following lectures were given in June 1993.

The Critical Theory Institute
John Carlos Rowe, Director

CONTENTS

Even a curriculum vitae is a kind of autobiography. Rudimentary and transparent though it is, it may reveal deeply personal traits. Certainly my own does; it makes abundantly clear that I have something of a problem with borders: in my peculiar psychic and intellectual economy borders are meant for crossing. More, they constitute irresistible lures. I seek them out—not to test their limits but to worry them, as a dog does a bone. Even as a working scientist, I found it hard to stay put, to keep from straying back and forth—in those days between biology and physics, between theory and experiment. And once I strayed beyond the borders of research science, shifted from doing science to writing about it, the problem only grew worse, for now I had many more boundaries to worry.

With this history I could hardly fail to welcome an invitation from the Critical Theory Institute, not only as an honor but as a challenge, a challenge to talk about science in the heartland of the "other culture." The chapters in this book, based on lectures given in June 1993, are the result—at one and the same time further instantiations of border crossing and, perhaps appropriately, *about* border crossings. They are about the traffic in normal scientific research between metaphors and machines, between software and hardware, between language and science—in short, about the normal processes of scientific exchange across that border between saying and doing, which scientists usually think of as steadfast and secure.

The notion that words are one thing, acts another, was radically undermined in 1955, when J. L. Austin laid out his theory of "speech-acts" in a series of lectures at Harvard University entitled *How to Do Things with Words* (1962). He argued that the function of language is not always or only descriptive; sometimes it is performative, hence the term *speech-act*. Unlike descriptive statements, speech-acts (his classic examples being bets, marriage pronouncements, and declarations of war) are not subject to tests of truth or falsity but need to be evaluated by a different criterion, by, say, their effectiveness (or, to use Austin's term, their *felicity*). As such, they are inherently and necessarily social, dependent on the existence of agreed-upon conventions about the effect of certain words, uttered in particular circumstances, by persons conventionally authorized to enact such effects.

Since Austin, the performative character of language has been extended by philosophers and literary theorists well beyond the domain of speech-acts. In keeping with these developments my assumption is that all language is performative and hence that all language, even scientific language, can and should be subjected to the criterion of efficacy. Efficacy is not invoked here in contradistinction to truth but in the pragmatist sense, as itself a potential measure of truth value; it also provides a way of grounding the social dependency of truth in material reality. It needs to be said, of course, that descriptive statements are performative in a rather different sense from that of speech-acts: not by virtue of directly enacting their referents but by their purchase on the ways in which we structure and construct our social and material worlds. In part, their force derives from the very impossibility of pure denotation. As the ubiquitous presence of metaphor attests, the classic distinction between literal and metaphoric holds no better in scientific than it does in ordinary language. Some of the force of descriptive statements, then, derives from the role of metaphor in constituting similarity and difference, in defining the "family resemblances" that form the bases on which we categorize natural phenomena (see, for example, Rosch and Mervis 1975, Hesse 1988) and in motivating the performance of particular experiments or the construction of particular technical devices.

Needless to say, not all metaphors are equally useful or, for that matter, equally captivating. The effectiveness of a metaphor, like that of a speech-act, depends on shared social conventions and, perhaps especially, on the authority conventionally granted to those who use it. It also depends on other family resemblances already in place. Consider, for example, the ways in which the process of biological fertilization has been figured. Twenty years ago that process could effectively and acceptably be described in terms evocative of the Sleeping Beauty myth (for example, penetration, vanquishing, or awakening of the egg by the sperm) precisely because of the consonance of that image with prevailing sexual stereotypes (see Martin 1991). Today a different metaphor has come to seem more useful and clearly more acceptable: in contemporary textbooks fertilization is more likely to be cast in the language of equal opportunity (defined, for example, as "the process by which egg and sperm find each other and fuse" [Alberts et al. 1990:868]). What was a socially effective metaphor twenty years ago has ceased to be so, in large part because of the dramatic transformation in ideologies of gender that has taken place in the interim.[1]

1. One facet of the cultural revolution of the last two decades is the dramatic increase in the numbers of women scientists, especially in the biological sciences. But the mere presence of women scientists is in itself no guarantor of a change in what counts as a socially acceptable metaphor. The shift in ideology, in cultural ideals of "masculine" and "feminine," was crucial, although that shift has of course been aided by the change in women's roles, just as the change in acceptable roles for women was, in turn, aided by a shift in the prevailing ideology of gender.

But what about the scientific effectiveness of a metaphor? Are not some metaphors more cognitively and technologically productive than others? Undoubtedly they are. But, perhaps more interestingly, they may be productive of different effects. In this example scientific productivity would have to be granted to both metaphors. One led to intensive investigation of the molecular mechanisms of sperm activity (yielding chemical and mechanical accsounts for the motility of sperm, for their adhesion to the cell membrane, and for their ability to effect membrane fusion), whereas the other fostered research permitting the elucidation of mechanisms by which the egg would have to be said to be active (for example, its production of the proteins or molecules responsible for both enabling and preventing adhesion and penetration).

Of course, the particular effectiveness of scientific metaphors depends not only on available social resources but also on the technical and natural resources that are available. Language does not simply construct reality. The productivity of the Sleeping Beauty metaphor, like that of the image of equal opportunity, required the cooperativeness of the material at hand—it required that evidence of active processes in either or both egg and sperm be elicitable, when sought, for the telling of one narrative or the other. It also required the availability of the sorts of technical apparatus that would be capable of recording such evidence and for providing the elements of a narrative plot.

I offer this story as the simplest example I can think of to illustrate a basic moral about the role of language in science. Yet simple as it is, it also suggests some of the historical and philosophical difficulties in charting the performative effects of metaphors in science. *Some* relation between the shift in metaphor in these accounts, the emergence of new research agendas, and the concurrent social transformation seems undeniable, but just how strong is the relationship? And is the account (the metanarrative) I have given here the only account of these events that can be given? What about the development of new technologies for representing the mechanisms of fertilization? Or what about the voracious (and insatiable) appetite of scientific researchers for new phenomena? Surely these must figure as well. If this story about saying and doing is to serve as history rather than as moral tale, it should be evident that a great deal more needs to be done to sort out the complex lines of influence and interactions of cultural norms, metaphor, and technical development.

My first chapter in this volume, "Language and Science: Genetics, Embryology, and the Discourse of Gene Action," is similarly incomplete. It tells a somewhat parallel story, not about egg and sperm but about cytoplasm and nucleus, about genes and organisms. The guiding metaphor of the discourse of gene action is one of genes as active agents, capable not only of animating the organism but of enacting its construction—as the physicist Erwin Schroedinger put it, as "law-code and executive power— . . . architect's plan and builder's craft—in one" (1944:23).

This two-sided image of the gene, part physicist's atom and part Platonic soul, was immensely productive for geneticists, both technically and politically. Of this there can be no doubt. For decades it permitted the framing of their questions in terms that allowed for a remarkably prolific research program—guiding their search for the "chains of reactions" (as Sturtevant put it) by which "genes produce their effects" and leading ultimately to the astonishing simplicity offered by modern molecular biology. Inevitably, of course, this way of talking about genes also had its costs, and these costs were felt most obviously by embryologists. The very glow of the geneticists' spotlight cast a deep and debilitating shadow on the questions, on the methods, indeed, on the very subject of embryology. It allowed neither time nor space in which the rest of the organism, the surplus economy of the soma, could exert its effects. What is specifically eclipsed in the discourse of gene action is the cytoplasmic body, marked simultaneously by gender, by international conflict, and by disciplinary politics. The virtual absence of this body from the research agenda of mid-twentieth-century biology in the United States cannot be explained simply by the absence of adequate technical facility but rather, as this story is intended to illustrate, requires us to understand how technical development itself builds upon such interlocking political dynamics. In short, it requires us to understand the ways in which scientific technique is both contributor to and product of discourse.

The image of gene as homunculus, as Lacan's "little man within a man," recurs in my discussion of the problem that life has historically posed for the second law of thermodynamics. In the second chapter, "Molecules, Messages, and Memory: Life and the Second Law," I trace a lineage from Maxwell's "Demon," introduced in the latter part of the nineteenth century, to the notion of code-script, which Schroedinger introduced in his famous (and purportedly vastly influential) book, *What Is Life?* (1944). Here I argue that Schroedinger's solution to the problem of life and the second law depends on locating a version of Maxwell's Demon in the molecular structure of the gene. As Schroedinger himself observes, genes are no ordinary molecules: they must contain not only the *plan* for executing the development of the organism but also must "somehow contain the means of putting [this plan] into operation." What is it, then, that gives the gene such remarkable powers? Schroedinger's answer (and his solution to the problem of life and the second law) is achieved, I suggest, by all but explicitly investing the gene with the power and the permanence of thought. In the last analysis it is the "cogito" that secures the continuance of life against the rages of time and, for Schroedinger in particular, against the rages of war.

The final chapter, "The Body of a New Machine: Situating the Organism Between Telegraphs and Computers," is intended to complicate the story I tell in the first chapter. It focuses more directly on the traffic between metaphors and machines and on the transformative effects this traffic has on the very terms of conventional social or technical histories of science. My first chapter, read either as social or as technical history, tells a salutary story of progress: after decades of occlusion, the subject of embryogenesis has been restored to biology, the discourse of "gene action" has been supplanted by a more adequate language of "gene activation," and the complexity and agency of the organismic body is finally being accorded its due. But the last chapter should go some way toward undermining whatever pleasure we might have drawn from that story's moral. In the process of its restoration the "body" of modern developmental biology has been radically transformed. It is not only that we now have different ways of talking of the body (for example, as a computer, an information-processing network, or a multiple input–multiple output transducer) but that, because of the advent of the modern computer (and other new technologies), we now have dramatically new ways of experiencing and interacting with that body. As a result of the technologies developed to elucidate that most elusive of nature's secrets, the beginnings of life, the material subject of embryology appears before biological researchers as a multimedia spectacle—visually available to a degree unthinkable in earlier years and tangibly and electronically available. This body is not only evocative of new ways of thinking, talking, and doing but, by virtue of the very techniques that have brought its microstructure into view (such as the gene tagging and fluorescent labels introduced to make it visible), it has already been constitutively transformed. The body of modern developmental biology is already a new kind of body; it is already "the body of a new machine."

If I may be permitted one last reflexive turn, it is worth pointing out that this book is also about trafficking across disciplinary borders—in the first chapter, between genetics and embryology; in the second, between physics and biology; and in the third, between cyberscience and molecular biology. These borders are of course all internal to the natural sciences and hence of considerably less institutional, intellectual, and moral concern than that more infamous boundary between the sciences and the humanities. Nonetheless, they illustrate many of the same issues. In particular they illustrate both the risks and the opportunities of disciplinary transgression. The accounts of interdisciplinary traffic offered in these chapters can perhaps best be made sense of if we think of disciplines as harboring intellectual and institutional resources. That would account for the extent to which their borders are policed, for the trading (and raiding) across disciplines that nonetheless persists, and even for the periodic attempts at annexation with which all histories of interdisciplinary activity are rife.

It might not be inappropriate to bring this moral a bit closer to home and to at least register a question about the opportunities and risks of trading, as I do in these chapters, across that most notorious and most vigilantly disciplined of all academic boundaries, across the two-culture divide. What are the intellectual and institutional resources to be risked or to be gained by the borrowing of literary techniques for the study of science and, conversely, by the addition of scientific texts to the mill of literary criticism? This question, I suggest, requires our collective consideration; indeed, in view of the particular urgency it has begun to assume in the wake of the current academic wars, it might be more accurate to say that it demands such consideration. The viability of such ventures in border crossing as these lectures may in fact depend on it.

REFIGURING LIFE

I

LANGUAGE AND SCIENCE: GENETICS, EMBRYOLOGY, AND THE DISCOURSE OF GENE ACTION

Adapted and expanded from my Tanner lecture,
"Rethinking the Meaning of Genetic Determinism,"
delivered at the University of Utah, February 18, 1993
(Keller 1994). This essay also appears in *Great Ideas
Today* 1994:2–29 (Chicago: Encyclopaedia
Brittanica, Inc.).

✴

A belief long standing among geneticists (and one that has acquired greater currency in recent years for the public at large) is that genes are the primary agents of life: they are the fundamental units of biological analysis; they cause the development of biological traits; and the ultimate goal of biological science is the understanding of how they act. Such confidence in the power and agency of genes—codified in what I call "the discourse of gene action"—has been of immense importance to the history of genetics and, most recently, to the launching of the Human Genome Initiative. But what does attributing (or for that matter, denying) causal power to genes mean? To what extent does this way of talking reflect a set of "natural facts," and to what extent does it reflect the facts of a particular disciplinary culture? And is it just a way of talking? Is it not also a way of thinking, a way of seeing, and a way of doing science? These are some of the questions I want to consider in the context of the history of two major areas of twentieth-century life science: genetics and embryology.

Historians of biology routinely note that for nineteenth-century biologists, the term *heredity* referred to both the "transmission of potentialities during reproduction *and* [the] development of these potentialities into specific adult traits" (Allen 1985:114). The question that compelled their interest above all others was, as August Weismann put it in 1883, "How is a single germ cell capable of reproducing the entire body with all its details?" (quoted in Sander 1986:363).

At the turn of the century, however, a crucial change began to occur. In 1900 Mendel was rediscovered; in 1902 Mendelian "factors" were tied to chromosomal structures; the term *genetics* was coined in 1905 and *gene* in 1909. In 1915 T. H. Morgan published *The Mechanism of Mendelian Heredity*; in 1916 the first genetics journal was founded, clearly marking the fact that a new discipline was off and running. For the new geneticists, the distinction between *genotype* and *phenotype* (introduced by Wilhelm Johannsen in 1911) provided a useful lexicon for distinguishing the problem of hereditary transmission from that of embryonic development (see Allen 1986: 127–28 and Sapp 1987:49). Indeed, the emergence of genetics coincided with the redefinition of the term *heredity* to refer exclusively to transmission: what had previously been seen as two aspects of a single subject (transmission and development) came to be regarded (at least by Americans) as distinct concerns. By the early decades of the twentieth century, the study of transmission had become the province of genetics, whereas that of development—now split off from genetics—continued as the province of embryology.

Two separate disciplines, with two different sets of concerns, emerged. In a passage from his 1926 book, *The Theory of the Gene*, T. H. Morgan described their relation as follows:

> Between the characters, that furnish the data for the theory and the postulated genes, to which the characters are referred, lies the whole field of embryonic development. The theory of the gene, as here formulated, states nothing with respect to the way in which the genes are connected with the end-product or character. The absence of information relating to this interval does not mean that the process of embryonic development is not of interest for genetics . . . but the fact remains that the sorting out of the characters in successive generations can be explained at present without reference to the way in which the gene affects the developmental process. (1926a:26)

Elsewhere, the same year, he cautioned that

> the confusion that is met with sometimes in the literature has resulted from a failure to keep apart the phenomenon of heredity, that deals with the transmission of the

hereditary units, and the phenomena of embryonic development that take place almost exclusively by changes in the cytoplasm. (1926b:490)

Genetics, we must remember, was still a relatively new discipline, struggling to establish itself against the established hegemony of embryology and physiology. The "phenomena of embryonic development" and their concurrent "changes in the cytoplasm" continued as powerful draws on the attention and interest of most biologists. Just two years earlier, in 1924, the German embryologist Hans Spemann and his graduate student Hilde Mangold (née Proscholdt) had published their classic paper on the "organiser," thereby sparking a tremendous burst of interest in the causal dynamics of embryonic development.[1] Certainly, genetics provided a powerful methodology for tracking the transmission of differences among existing organisms, but it had no answer to how a single germ cell might produce an organism. The pursuit of that question continued as the province of embryology.

Yet even in the early days of genetics, when the gene was still merely an abstract concept and the necessity of nuclear-cytoplasmic interactions was clearly understood, geneticists of Morgan's school tended to assume that these hypothetical

1. Hilde Mangold died in an accident just months before the paper was published.

particles, the genes, must somehow lie at the root of development. If in some of his writings Morgan gave the impression of being ecumenical, granting to embryology a separate but equal disciplinary status and a separate but equal object of study (the cytoplasm), at other times he was quite clear about the proper epistemological ordering of the two disciplines. Although he well recognized that geneticists could say nothing about what genes are, or how they were subsequently connected to the formation of adult characters or traits, and little about how they interacted with the cytoplasm of the fertilized egg (the specifically maternal contribution), he nonetheless wrote in 1924, "It is clear that whatever the cytoplasm contributes to development is almost entirely under the influence of the genes carried by the chromosome, and therefore may in a sense be said to be indifferent" (728). Others went even further. In an attempt to clarify Morgan's position, the geneticist R. A. Brink explained that "the Mendelian theory postulates discrete, self-perpetuating, stable bodies—the genes—resident in the chromosomes, as the hereditary materials. *This means, of course, that the genes are the primary internal agents controlling development*" [emphasis added]. Brink described the great advantage of genetics over other approaches as follows:

With the primary internal mechanism resolved into definite units which may be combined in various groups, . . . the hereditary complex need no longer serve merely as the passive object in physiological experimentation but may itself be varied in a precise fashion. . . . We are now in a favorable position to get at the dynamic properties of the hereditary mechanism by means of an analysis of the action of its separate elements. This, it seems to us, is the signal contribution which genetics makes to our outlook upon the problems of developmental physiology. (1927:281–82)

To H. J. Muller, Morgan's student, the most remarkable characteristic of the gene was that it possesses the property he called "specific autocatalysis" (by which he meant self-replication). "Still more remarkable," he wrote, "the gene can mutate without losing its specific autocatalytic power" (1926:200). Largely for this reason, he entitled his own 1926 paper "The Gene as the Basis of Life" (it is said that he refused to change his title to "The Gene as a Basis for Life"). In it he concluded that

the great bulk . . . of the protoplasm [is], after all, only a by-product of the action of the gene material; its "function" (its survival-value) lies only in its fostering the genes, and the primary secrets common to all life lie further back, in the gene material itself." (200–201)

Today it may be hard to see what could be controversial in such claims. The attribution of agency, autonomy, and causal primacy to genes has become so familiar as to seem obvious, even self-evident. However, what I want to do is attempt to dislodge that familiarity—by citing these arguments in their historical context (using the now somewhat quaint language in which they were first posed) to make it possible to see them as novel and, thereby, to see something of the process by which they acquired their familiarity and ring of truth.

Language in Action

It is well known that by the mid-1920s, the taming of *Drosophila* and corn as model organisms for tracking the transmission of hereditary traits lent genetics a rigor and productivity that other disciplines could scarcely match. But the first generation of geneticists—Morgan and his school—did more than develop the techniques and practice of genetics as a rival to embryology; they also forged a way of talking about genes—about their role and meaning in reproduction, growth, and development. When Muller identified the gene as *the* basis of life, he was claiming for it both ontological and temporal priority. First the gene, then the remaining protoplasm (the cytoplasm), which appears as a *by-product*, the only function is of which is that of a facilitating environment, to *foster* the gene. First the gene, then life. Or, rather, with the gene comes life. The concept of gene invoked here is Janus faced: it is part physicist's atom and part Platonic soul—at one and the same time a fundamental building block and an animating force. Only the *action* of genes can initiate the complex manifold of processes comprising a living organism.

But what exactly is it that genes *do*? This of course neither Muller nor Brink nor Morgan could say. The notion of "gene action" may even have been facilitated by the very absence of knowledge of what a gene is (in the sense that not knowing what a gene is may have made it easier to attribute to it any, even miraculous, properties). But although these early geneticists could say nothing about the nature of the presumed source of all subsequent growth and development, could give no scientific account of gene action, they offered future generations of geneticists something equally valuable.

Scientists usually assume that only their data and theories matter for scientific progress, that how they talk about these data and theories does not matter, that it is irrelevant to their actual work. But in introducing this particular way of talking, the first generation of American geneticists provided a conceptual framework that was critically important for the future course of biological research. To capture both its rhetorical and conceptual force, I will call this way of talking the "discourse of gene action"—a discourse that was, for genetics, undeniably productive. It enabled geneticists to get on with their work without worrying about their lack of information about the nature of such action—to a considerable degree, it even obscured the need for such information. (Throughout the interwar period, American geneticists routinely invoked the notion of gene action as if its meaning was self-evident.) At the same time, the attribution of agency, autonomy, and causal responsibility to genes lent primacy both to the object of geneticists' concern and to the discipline of genetics—in their own eyes and in the eyes of others. They were dealing with *the* basis of life. If, as Brink wrote, the hereditary complex is elevated from a "passive object" to a locus of primary activity, the student of that hereditary complex is, by the same move, also elevated to primary activity.

Indeed, I suggest that the discourse of gene action provides the specific hallmark (or trademark) of the American school of Morganian genetics, especially of its approach to development. If its first use was to bracket the question of development, when a number of American geneticists did turn their attention to development in the mid-1930s, such a lexicon helped define the approach they took: it framed the questions they could or could not meaningfully ask, the organisms they would choose to study, experiments that did or did not make sense to do, the explanations that were or were not acceptable. In this sense, the discourse of gene action served cognitive as well as political functions. Ian Hacking has suggested that every scientific discipline has its own "style of reasoning," and that this "style of reasoning" constitutes the epistemological context of that science. In other words, a style creates the very possibility for truth or falsehood and therefore determines what counts as objective (1982). My notion of discourse is close to Hacking's notion of style.

Needless to say, most embryologists found this way of talking about the relation between genes and development—a way that recasts the dynamics of development as a consequence of gene action—markedly less congenial. It offered the student of development not a separate domain of inquiry (as Morgan's remarks implied) but a promissory note for inclusion or, more accurately, for incorporation. As early as 1924, Spemann, Morgan's most important counterpart, wrote,

> The previous progress [of genetics] has been amazing, and it is not from a feeling of futile labours but rather from being aware of their paramount powers of appropriation that geneticists now are on the look-out for new connexions. They have cast their eye on us, on *Entwicklungsmechanik*. (293)

A decade later, Ross Harrison sounded a similar warning in his presidential address to the American Association for the Advancement of Science:

Now that . . . the "Wanderlust" of geneticists is beginning to urge them in our direction, it may not be inappropriate to point out a danger in this threatened invasion.

The prestige of success enjoyed by the gene theory might easily become a hindrance to the understanding of development by directing our attention solely to the genome. . . . Already we have theories that refer the processes of development to genic action and regard the whole performance as no more than the realization of the potencies of the genes. Such theories are altogether too one-sided. (1937:372)

Embryologists had good grounds for concern. Not only was the status of their discipline under threat; so too was the status of their question: how *does* a germ cell develop into a multicellular organism? If the genetic content of all cells in an organism is the same, how is it possible to make sense of the emergence of the manifest differences among all the cells that make up a complex organism? To the embryologists it seemed self-evident that this problem of differentiation, so deeply at the heart of their own concerns, was simply incompatible with the notion that the gene was the exclusive locus of action.[2] As Morgan himself (speaking now as an embryologist) subsequently acknowledged,

2. Geneticists could study only variations in already existing organisms; how organisms come to be formed in the first place was thus beyond their ken.

The implication in most genetic interpretation is that all the genes are acting all the time in the same way. This would leave unexplained why some cells of the embryo develop in one way, some in another, if the genes are the only agents in the results. (1934:9)

Few if any geneticists heeded Morgan's warning. (Even Morgan did not heed his own warning.) Instead, those interested in the relation between genes and development found another route: they changed the subject. Or, more precisely, they transformed the embryologist's question into a different one. Alfred H. Sturtevant spelled out how to do this. He opened his paper on the developmental effects of genes at the 1932 International Congress of Genetics by observing,

One of the central problems of biology is that of differentiation—how does an egg develop into a complex many-celled organism? That is, of course, the traditional major problem of embryology; but it also appears in genetics in the form of the question, "How do genes produce their effects?" (304)

To address this question, he examined the correlation of eye pigment with gonad constitution in *Drosophila* gynandromorphs and concluded,

It is clear that in most cases there is a chain of reactions between the direct activity of a gene and the end-product that the geneticist deals with as a character. . . . The type of experiment that I have described may be considered as a beginning in the analysis of certain chains of reactions into their individual links. (307)

I do not want to underestimate the importance of Sturtevant's suggestion that it was possible to use spontaneously occurring mosaics in *Drosophila* for the study of development, especially because *Drosophila* had, until then, been thought to be beyond the reach of embryology.[3] I mean merely to underscore the linear format in which he posed the problem (a gene "produces its effect" through a "chain of reactions," or "direct activity" that produces the end-product, its character). Contrast this formulation with that of Richard Goldschmidt, the leading figure in German physiological genetics, who had also been studying cases of intersexuality in insects (mostly moths) for some time and who was also concerned with the nature of gene action. Many differences in style are conspicuous (to an American reader, Goldschmidt was typically grandiose, leaning always toward overarching generalization),[4] but the differ-

3. In fact, forty years later Arturo Garcia-Bellida and John Merriam (1969) resurrected a drawerful of Sturtevant's actual data to produce a fate map for *Drosophila*.

4. In part, this tendency reflects the difference in "national styles" that Jane Maienschein has described. In a letter to L. C. Dunn (in response to Dunn's frank criticism of his undue quickness to generalize), Goldschmidt wrote, "I certainly realize my weakness to run ahead of the facts with conclusions; this is derived from my genetic makeup which forces me to assign to new facts immediately their place within the whole in order to satisfy my sense of 'national normalness' " (May 27, 1940).

ence I want to point out is more subtle: Goldschmidt's emphasis was on *systems* of coordinated reactions of particular velocities ("Development ought to be disentangled into a series of coordinated reactions of definite velocities" [1932:345]); his search was precisely for the dynamic properties of such systems (*Zusammenspiel der Reaktionen*). To Goldschmidt gene action meant that genes were both catalysts and catalyzed, actors and "reacting substances" (343). In his 1927 book, he had advocated a "scheme of 'substrate-induced' differential gene activation" where "gene activation" (*Genaktivierung*) is effected by the appearance of an appropriate substrate.[5]

To American geneticists, however, Sturtevant's rephrasing of the problem—and his recommendation to geneticists of the task of analyzing the "chains of reactions into their individual links"—had an immense appeal; certainly, it seemed vastly clearer than Goldschmidt's formulations. As they no doubt intuited, Sturtevant's rephrasing accomplished a great deal. Once the problem of development is translated into the question of how genes produce their effects, the task is immediately—and almost miraculously—simplified. No longer need a geneticist become bogged down in the complex dynamics of eggs and multicellular organisms; studying

5. In a similar vein Goldschmidt speculated that "the nurse cells are responsible for polarity and/or gradients in the insect egg cell" (Sander 1986:368–69).

single-celled organisms, which provide a better opportunity to analyze "chains of reaction," ought to suffice. George Beadle and Arthur Tatum chose *Neurospora*, a single-celled organism that can be cultured in vitro, and their choice paid off handsomely. In 1940 they proposed their explanation of how genes produce their effects in a form that came to be known as the "one gene–one enzyme" hypothesis. Beadle and Tatum provided a particular kind of answer to the question of how a gene produces its effects—namely, it catalyzes a specific chemical reaction. More colloquially, it makes an enzyme. At last the mysterious notion of gene action seemed to have real content. With the one gene–one enzyme hypothesis, developmental genetics could henceforth be understood as the biochemistry of gene action.

Together, turning to *Neurospora* and to the biochemistry of gene action proved to be of decisive importance to the future development of genetics. The shift provided critical encouragement for the development of bacterial genetics and, eventually, of molecular biology. The rest of the story is familiar. In 1953, with the definitive identification of DNA as the genetic material, J. D. Watson and Francis Crick struck gold. Simple hydrogen bonding provided the secret of how genes reproduce themselves, and nucleic acid sequences revealed how they make enzymes. As Watson and Crick discreetly wrote, "In a long molecule, many different permutations are possible, and it therefore seems likely that the precise sequence of the bases is the code which carries the genetical information" (1953:967). All one needed to know was the code, and soon that was forthcoming as well.

Geneticists and molecular biologists were euphoric: there, surely, must be the answer! DNA carries the "genetical information" (or program), and genes "produce their effects" by providing the "instructions" for protein synthesis. DNA makes RNA, RNA makes proteins, and proteins make us. It was, without doubt, one of the greatest milestones in the history of science. But still, one might ask (although few people did at the time), what kind of answer is this? What, in fact, do *information, program, instruction* or even the verb *makes* actually mean?

Watson and Crick have gotten a lot of credit for their work and deservedly so, but one contribution has, I fear, been overlooked: their introduction of the information metaphor to the repertoire of biological discourse was a stroke of genius. The story of this metaphor—its uses and implications—is immensely rich and has been extensively explored by others (see especially Doyle 1993 and Kay, forthcoming) but perhaps a few brief comments might nonetheless be in order. Just a few years earlier, the mathematician Claude Shannon had proposed a precise quantitative measure of the complexity of linear codes. He called this measure *information*—by design, the term was independent of meaning or function—and by the early fifties, "information theory" had become a hot subject in the world of communications systems. It seemed to hold enormous promise for analyzing all sorts of complex systems, even biological systems. Because DNA seemed to function as a linear code, using this notion of information for genetics appeared to be a natural. But as early as 1952, geneticists recognized that the technical definition of *information* simply could not serve for biological information (because it would assign the same amount of information to the DNA of a functioning organism as to a mutant form, however disabling that mutation was). Thus the notion of genetical information that Watson and Crick invoked was not literal but metaphoric. But it was extremely powerful. Although it permitted no quantitative measure, it authorized the expectation—anticipated in the notion of gene action—that biological information does not increase in the course of development: it is already fully contained in the genome. This move and, even more, the collapsing of *information* with *program* and *instruction* vastly fortified the concept of gene action. Just as Erwin Schroedinger had anticipated in his widely read work, *What Is Life?*, the "chromosome structures are law-code and executive power—or, to use another simile, they are architect's plan and builder's craft—in one" (1944:23).[6]

6. For a particularly elegant elaboration of this point, see Doyle (1993).

Classical embryologists would surely not have been happy with this turn of events—their questions, their organisms (even the lowly *Drosophila* had come to be seen as too complex, too messy), and they themselves had been left behind—but a new generation of biologists had little cause to look back. The first generation of molecular biologists could not say how an egg turns into an organism (in other words, they could say nothing about how a gene comes to make the particular enzymes that are needed for the development of a many celled organism, in the right amounts, at the right time, and in the right place), but they had a powerful new rhetorical resource for managing such questions. They could talk instead about development in the abstract and the genetic programs or instructions that are needed to guide it. In his presidential address to the British Association for the Advancement of Science in 1965, Sir Peter Medawar offered something of a retrospective eulogy to embryology:

> Wise after the event, we can now see that embryology simply did not have, and could not have created, the background of genetical reasoning which would have made it possible to formulate a theory of development. . . . Embryonic development . . . [must] be an unfolding of pre-existing capabilities, an acting-out of genetically encoded instructions. (1328–29)

Twenty years later, the progression from Watson and Crick to the Human Genome Initiative, as Watson himself has so often reminded us, appeared straightforward and logical. If all development is merely an unfolding of preexisting instructions encoded in the nucleotide sequences of DNA—if our genes make us what we are—it makes perfect sense to set the identification of these sequences as the primary and, indeed ultimate, goal of biology.

An Irony of History

What then do I mean when I say that the discourse of gene action—now augmented with metaphors of information and instruction—exerted a critical force on the course of biological research? Can words have force in and of themselves? Of course not. They acquire force only through their influence on human actors. Through their influence on scientists, administrators, and funding agencies, they provide powerful rationales and incentives for mobilizing resources, for identifying particular research agendas, for focusing our scientific energies and attention in particular directions. The discourse of gene action has worked in just these ways. And it would be foolhardy to pretend it has not worked well. The history of twentieth-century biology is a history of extraordinary success; genetics—first classical, then molecular—has yielded some of the greatest triumphs of modern science. Indeed, this way of talking has proved so powerful that now, after all these years, it seems to be finally on the verge of making good the promissory note that Morgan and his school extended in the early part of the century—and not just rhetorically but in actual scientific practice. During the last few years, molecular biology has made extraordinary progress in elucidating just how it is that (as they say) genes control development.

But a funny thing happened on the way to the holy grail. That extraordinary progress has become less and less describable within the discourse that fostered it. The dogmatic focus on gene action called forth a dazzling armamentarium of new techniques for analyzing the behavior of distinct gene segments, and the information yielded by those techniques is now radically subverting the doctrine of the gene as sole (or even primary) agent. It has also become conspicuously evident that there were all along serious problems with the discourse of gene action—in addition to its productive blindness to questions of development and cell differentiation. As Richard Lewontin reminds us,

> DNA is a dead molecule, among the most nonreactive, chemically inert molecules in the world. . . . [It] has no power to reproduce itself. Rather it is produced out of elementary materials by a complex cellular machinery of proteins. While it is often said that DNA produces pro-

teins, in fact proteins (enzymes) produce DNA. The newly manufactured DNA is certainly a *copy* of the old, . . . but we do not describe the Eastman Kodak factory as a place of self-reproduction [of photographs]. (1992:33)

He continues,

Not only is DNA incapable of making copies of itself, . . . but it is incapable of "making" anything else. The linear sequence of nucleotides in DNA is used by the machinery of the cell to determine what sequence of amino acids is to be built into a protein, and to determine when and where the protein is to be made. But the proteins of the cell are made by other proteins, and without that protein-forming machinery *nothing* can be made. There is an appearance here of infinite regress . . . , but this appearance is an artifact of another error of vulgar biology, that it is only the genes that are passed from parent to offspring. In fact, an egg, before fertilization, contains a complete apparatus of production deposited there in the course of its cellular development. We inherit not only genes made of DNA but an intricate structure of cellular machinery made up of proteins. (33)

But of course, the reader might think, didn't we know this all along? Well, yes and no. Yes in the sense that, apart from the reference to DNA, it is the sort of observation embryologists used to make all the time. But no in the sense that, except for an occasional aside (such as Morgan's), geneticists did not; interestingly, Lewontin is a geneticist, not an embryologist. The simple fact is that for many years geneticists had little reason to refer to eggs and their cytoplasmic structure and even less reason to talk about events before fertilization. The discourse of gene action had established a spatial map that lent the cytoplasm scientific invisibility (at least to geneticists—"indifferent" was how Morgan described the cytoplasm) and a temporal map that defined the moment of fertilization as origin, with no meaningful time before fertilization. This schema offered neither time nor place in which to conceive of the egg's cytoplasm as exerting *its* effects.

With the emergence of molecular biology in the 1950s and 1960s, and its powerful metaphors of information and programs, the significance of the cytoplasm eroded even further. And once the bacterium *E. coli* came to serve as the model organism (recall Jacques Monod's famous remark, "What's true for *E. coli* is true for the elephant" [cited in Judson 1979:613]), questions about eggs and fertilization, or about developmental processes of any kind, ceased to have any applicability. What is new is that Lewontin's commonplace observations have once again come to make sound biological sense, even in genetics. Genetics itself has turned once again to the study of higher organisms—of flies, frogs, and mice—and with that return has rediscovered all the old problems of embryogenesis. Indeed, one might say that the long-awaited rapprochement of genetics and embryology has finally occurred. But the terms of that rapprochement turn out to be vastly different from those imagined by the first generations of geneticists and molecular biologists. Current research—drawing on the phenomenal technical successes of molecular biology, and even on the sequence information emerging from the Human Genome Initiative—invites (ever more insistently) a shift in locution in which the cytoplasm is just as likely as the genome to be cast as the locus of control. How did this happen?

Tentative signs of this transformation—some of them apparently quite innocuous—may have appeared as early as the mid-1960s. Just when Medawar (and others) were singing eulogies to embryology, a new—and more agnostic—term had already been coined: *developmental biology.* For those still interested in problems of embryogenesis, the renaming of their field clearly represented a strategy of accommodation. Not only had the term *embryology* become hopelessly tainted, so too had such kindred terms as *growth.*[7] On the other hand, for newcomers from molecular

7. In 1964 the Society for the Study of Growth and Development, commonly known as the Growth Society, officially renamed itself as the Society for (the Study of) Developmental Biology. Clement Markert, then president of the society, suggests that the change in name "was motivated by two reasons: (1) the Growth Society had declined somewhat so that it did not have a very good image; and (2) and more important, the term 'growth' was not descriptive of the Society. . . . The term [developmental biology] was much more descriptive than any previously used term, such as growth or embryology, and did, in fact, enhance the scientific image of the Society in an appropriate fashion" (Clement Markert, letter to author, February 14, 1992).

biology, the term *developmental biology* had a useful ring of universality—it was free of the metazoan bias explicit in the term *embryology.*

In any case, the new name took and, by the late 1960s, so did its practice. In 1968 Eric Davidson published his influential *Gene Activity in Early Development,* a work written for molecular biologists that focused squarely on the issue of differentiation in embryogenesis. Two years later the first Gordon Conference in Developmental Biology was convened, and in 1972 the first graduate programs explicitly designating developmental biology appear in *Peterson's Annual Guide to Graduate Programs in the Biological Sciences.* By 1981 the number of such programs had grown to fifty, by 1992 to eighty, and in 1985 the Stanford University Board of Trustees voted to establish an entire department in developmental biology.

Nonetheless, the rhetorical shift from gene action to gene activation was neither immediate nor uniform. As recently as 1984 David Baltimore was still invoking the more familiar language of molecular biology to explain the distinction between modern genetics and classical physiology (or embryology):

> The approach of genetics . . . is to ask about blueprints, not machines; about decisions, not mechanics; about information and history. In the factory analogy, genetics leaves the greasy machines and goes to the executive

suite, where it analyzes the planners, the decision makers, the computers, the historic records. : . . . Biologists needed to find the cell's brain. (150)

Seven years later, however, Baltimore wrote (with Helen Blau) of the extent to which differentiation is governed by "active control" mechanisms, in which "the expression state of each gene [is] determined by the dynamic interaction of regulatory proteins present in the cell at any given time" (1991:781).

Indeed, even as Baltimore was speaking of the need to find the cell's brain in "the executive suite" (the DNA), the cell's brain was already moving out of the executive suite and into the factory. In 1984 Sidney Brenner, himself one of the major architects of molecular biology, acknowledged:

At the beginning it was said that the answer to the understanding of development was going to come from a knowledge of molecular mechanisms of gene control. . . . I don't know if anyone believes that anymore. The molecular mechanisms look boringly simple, and they don't tell us what we want to know. We have to try to discover the principles of organization, how lots of things are put together in the same place. I don't think these principles will be embodied in a simple chemical device, as it is for the genetic code. (quoted in Lewin 1984)

Today geneticists see the really "smart genes" as those that have the capacity to respond to a complex of signals encoded in cytoplasmic proteins. Genes may be "smart," but "the brain of the smart gene," is not to be found in the genes themselves: as Eric Davidson puts it, it is a "complicated assemblage of proteins known as a transcription complex" (quoted in Beardsley 1991:87). The point is that as we learn more about how genes actually work in complex organisms, talk about "gene action" subtly transmutes into talk about "gene activation," with the locus of control shifting from genes themselves to the complex biochemical dynamics (protein-protein and protein-nucleic acid interactions) of cells in constant communication with each other. *Scientific American* glosses this shift as the "news" that "organisms control most of their genes" (Beardsley 1991:87).

New metaphors abound. Marking the long-overlooked distinction between program and data, Henri Atlan and Moshe Koppel suggest "an alternative metaphor of DNA as data to a parallel computing network embedded in the global geometrical and biochemical structure of the cell"(1990:335). H. F. Nijhout proposes a yet more radical inversion. In lieu of the metaphors of control and 'programs that have so pervaded modern thinking in molecular, developmental, and evolutionary biology—and that, he says, "have shaped priorities in research"—Nijhout suggests that "a more balanced, and useful view of the role of genes in development is that they act as suppliers of the material needs of development and . . . as context-dependent catalysts of cellular changes" (1990:441). Genes, he concludes,

> are passive sources of materials upon which a cell can draw, and are part of an evolved mechanism that allows organisms, their tissues and their cells to be independent of their environment by providing the means of synthesizing, importing, or structuring the substances (not just gene products, but all substances) required for metabolism, growth and differentiation. The function of regulatory genes is ultimately no different from that of structural genes, in that they simply provide efficient ways of ensuring that the required materials are supplied at the right time and place. (444)

Nijhout's proposal may be extreme. But there is no question that a new way of talking is in the air, in keeping with the emergence of a new biology: molecular biologists, it appears, have "discovered the organism." The new developmental biology brings with it a resurgence of interest in many of the problems of organization and morphogenesis that had occupied an earlier generation of embryologists and even a resurrection of a number of the same experimental protocols. The findings that result point neither to cytoplasmic nor to nuclear determination but rather to a complex but highly coordinated system of regulatory dynamics that operate simultaneously at all levels: at the level of transcription activation, of translation, of protein activation, and of intercellular communication—in the nucleus, in the cytoplasm, indeed in the organism as a whole.

So what is it that I find so interesting about this story? For a scientist (even for a semi-lapsed scientist such as me), what compels the greatest interest must surely be the specific content of the conceptual revolution now under way. The shift in discourse we are now seeing in the literature marks a conceptual shift of startling magnitude; it will require us to learn how to think in radically new ways.

Sixty years ago men such as Joseph Needham, C. H. Waddington, and J. H. Woodger sought a language for the complex dynamics relating nuclear and cytoplasmic elements in the process philosophy of Alfred North Whitehead. Today Whitehead's language is too foreign to us to be of use. But we have other resources to compensate—especially in mathematics and computers. For some time now, a number of workers—Stuart Kauffman and René Thomas, among others—have been developing models for genetic networks that represent a great advance over more simplistic notions of gene action. These models illustrate how networks of genes in interaction can give rise to stable, self-perpetuating patterns of biochemical dynamics of a kind radically different from anything autonomously acting genes could ever yield. In so doing, they give substance to Waddington's earlier notions of epigenetic pathways and, at the same time, automatically and irrevocably undermine traditional divisions between genetic and epigenetic. But as interesting as they are, such models are only a beginning. In keeping with the new talk of cytoplasmic control, it would also be useful to develop models of somatic networks of interacting proteins, in which genes would be the covert intermediaries of protein interaction rather than proteins being the intermediaries of gene interaction (as they are in gene network models).

Ultimately, of course, one needs full-scale models of genes and proteins in interaction—of a kind that large-scale computers are now making possible. In the end, I think that the most important function of all these models will be to stimulate the growth of just those intuitions about interactive and emergent phenomena that past discourses have so helped to stymie. I have no doubt that the effect will be a transformation in the way we think about biological systems that will make the changes we have already begun to witness look like mere harbingers.

However, for a historian of science, this story provokes other questions. Put simply, they are twofold: what lent the discourse of gene action such persuasiveness for so many years? And why is it now giving way? (Or, we might ask why did embryology languish for so many years, and what has permitted its return today?) These are different versions of the same questions, just because of the extent to which the fate of embryology has historically been so intimately linked with talk of gene action. Posed either way, they are far more difficult to answer than naïve empiricism might suggest. The simplistic answers might go like this: embryology languished because it was bad and unproductive science; we talked about gene action because we didn't know better; indeed, developmental phenomena are so difficult to study that real progress was impossible until the advent of techniques of recombinant DNA that molecular biology has brought. All these claims might be true and still only part of the story. What they leave out is the entire issue of motivation.

Relatedly, they also ignore the awkward fact that the first experimental studies to spark the interest of molecular biologists in the early development of higher organisms relied solely on classical techniques that were labor intensive, to be sure, but that had long been available. I think especially of the studies of "maternal (or cytoplasmic) effect" mutants and of cytoplasmic rescue in *Drosophila* first undertaken by Alan Garen and others in the early seventies and carried to such dramatic fruition a few years later by Christianne Nüsslein-Volhard and her colleagues.[8] What these studies did was establish the critical role played by the cytoplasmic structure of the egg before fertilization—before time zero. The most conspicuous question is why these efforts were undertaken in the 1970s and not before. Time and space preclude my going into the details of these studies, but they reveal, as Garen and others confirm, that no technical impasse prevented their being done years, if not decades, earlier.[9] Maternal effect mutants—even in *Drosophila*—had been accumulating since the early part of the century; the most crucial technical instrument, the micromanipulator, had also been well developed and was in wide used by the 1930s. Of course, it might be argued that *Drosophila* was an exceedingly dif-

8. Nüsslein-Volhard began her studies with a mutant that had been isolated by Alice Bull more than ten years earlier. But as Nüsslein-Volhard explains, "At the time of Alice Bull, modern biologists were not interested in development and fly people were not interested in embryos" (Nüsslein-Volhard, letter to author, April 4, 1992).

9. In a telephone interview on April 29, 1992, I asked Alan Garen whether his 1972 experiment on cytoplasmic rescue in *Drosophila* could have been done earlier. The conversation proceeded as follows:

Garen: Certainly—all it took was a good glass blower.

EFK: Why wasn't it?

Garen: We didn't have maternal effect mutants.

EFK: But the mutant you used had actually been isolated by Sheila Counce in the mid fifties.

Garen: Well, yes. [pause] It's a good question. I don't know why, I guess no one was interested.

ficult organism to study embryologically (and it surely was), but even this ostensible impasse had been largely overcome by the early 1950s—again, by applying long-available techniques. What was missing—both from the study of *Drosophila* embryology and from the more specific examination of maternal effects—was the motivation to invest the necessary effort. And once again nomenclature is relevant: since 1930 geneticists had argued for the term *delayed inheritance* as a more accurate description of these mutants.[10] What relabeling *maternal effects* as *delayed inheritance* achieved was a reinforcement of the understanding of phenotypic expression, however late, as pure outcome—in other words, as merely epiphenomenal and hence lacking causal weight of its own. More generally, the belief—prevalent among geneticists at least since the mid twenties—that the genetic message of the zygote "produces" the organism, that the cytoplasm is merely a passive substrate, could not but sap the motivation needed

10. See Boycott et al. 1930:52. Another term for these effects, *maternal inheritance,* was equally misleading and, because it led to the bracketing of such effects with controversial arguments for cytoplasmic inheritance, contributed equally to the long disregard of maternal effects. As Sturtevant and Beadle put it in 1939, "Cases of this type are sometimes said to show *maternal inheritance*—a misleading term, since inheritance is perfectly normal, the unusual feature being that the phenotypic expression is delayed a generation. . . . In all these cases the characters described are clearly under genetic control; the only complication is that there is a delay in the expression of the phenotype" (329–31).

to undertake such undeniably difficult experiments.[11] The question therefore becomes what overcame that assumption.

If, as I have been arguing, the ways in which we talk about scientific objects are not simply determined by empirical evidence but rather actively influence the kind of evidence we seek (and hence are more likely to find), we must consider other factors if we are to understand the strength and persistence of the discourse of gene action. Let me, in the remainder of this essay, schematically indicate what some of these other factors were, at least as they operated between the two world wars.

11. The remarkable work of Nüsslein-Volhard surely attests to her independence and fortitude; it also provides a good example of what Harriet Zuckerman and Joshua Lederberg have described as "postmature discoveries" (1986). Another example of a similar delay might be seen in the recent successes in the analysis of insect metamorphosis. This work is generally regarded as beginning with a series of papers by Michael Ashburner in the mid-1970s on the use of ecdysone to induce chromosome puffs and proposing a model for the "coordination of gene activation." However, in a recent review (1990), Ashburner credits two German geneticists working in the 1950s, W. Beerman and H. J. Becker, with the basic ideas for both the work and the model. Yet another instance might be seen in the work of John Gerhart from the 1970s, when he undertook to repeat some key classical experiments (heretofore unknown to molecular biologists) on egg rotation in frogs. I suggest that such cases can be taken to illustrate the loss of technological momentum resulting from the discourse of gene action and, furthermore, that similar relations between discourse and practice might help to illuminate the phenomenon of "postmature discovery" more generally.

A Metaphor in Context

In the 1930s the Swiss embryologist Oscar Schotte liked to illustrate the relations between embryology and genetics with a sketch of two views of the cell: as perceived by the embryologist, the nucleus is small, but as perceived by the geneticist, it virtually fills the entire cell (Sander 1986). His sketch uses the nucleus and cytoplasm as tropes for the two disciplines—each lends to its object of study a size in proportion to its perceived self-importance. In like fashion, the two disciplines lent to each object, nucleus and cytoplasm, their own attributes of agency, autonomy, and power. As L. C. Dunn put it, "Genetics had to be a bit pushy in order to get itself established" (1959:319).

However, the nucleus and cytoplasm also came to stand as tropes for national importance, agency, and power. The nucleus was the domain in which American genetics staked its unique strengths, associated with American interests (and prowess), whereas the cytoplasm was associated with European, especially German, interests and prowess. German biologists were often explicit about what they saw as the attempt by American geneticists to appropriate the entire field. In 1926 V. Haecker described the field between genetics and development as the "no-man's land" of somatogenesis—"a border field which by us has been tilled for quite some time. . . . The Americans have taken no notice of this" (233). Goldschmidt registered a similar complaint, attribut-

ing American indifference to "the rise of a school of geneticists to whom biological knowledge apart from Mendelism
did not seem necessary, whereby they were entirely content
with knowing the work of the schools most closely akin to
their own approach" (translated by Sander 1985:389).[12] This
tension showed some signs of abating as memories of World
War I faded in the late 1920s and early 1930s, only to resur-
face with Hitler's rise to power. Spemann's fierce (Bismarck-
ian) nationalism was recalled, as were his occasional references to the experimental superiority of the hair of Aryan
babies or to the special intuitive gifts of German scientists.[13]
Even Goldschmidt was suspect, although himself a Jew
obliged to emigrate in 1936.[14] On August 23, 1939, the Sev-

12. Elsewhere Goldschmidt commented, "It is really too bad that Morgan
and his students . . . have got stuck in such a narrow interpretation of genetic
phenomena and oppose at all costs any new idea, especially a physiological
one. . . . I have discussed this at length with my dear friend Morgan, but he
insists that a thing [phenotype] has been explained once one has mapped a
corresponding Mendelian factor" (quoted in Harwood 1993:50).

13. The use of a baby's hair was critical to Spemann's experimental
manipulation of early embryos. For further information about his nationalism,
see Horder and Weindling (1986).

14. In 1959 Dunn recalled Goldschmidt's saying, "I know you think I've
been foolish to think that they would miss me, of Jewish birth, but after all,
remember that my family was settled in the Rhineland in the [thirteenth cen-
tury]. What am I, if not German?" Dunn goes on to suggest that Goldschmidt
"had the same illusions about Hitler . . . that a great many of the German aca-
demicians had. . . . He regarded himself as perfectly secure; he had a profes-
sorship, he was Geheimrat, he was everything that a German professor ought
to be, and in the Institute he was God himself" (Dunn oral history:377).

enth International Congress of Genetics convened in Edinburgh.[15] Goldschmidt did not attend, but forty-two of his colleagues from Germany did. The minutes from the meeting read:

> For a day and a half the congress was able to immerse itself in its own enjoyable affairs. It even danced. But on the evening of the 24th its serenity was shattered. War, that outmoded futility of irrational immaturity, the antithesis of everything we represented, was about to overwhelm us. Britishers in Germany had been advised to leave for home. The German delegation, therefore, had no choice but to do likewise. . . . We had met as geneticists sharing the same interests and enthusiasms; suddenly we were required to behave as nationals with fiercely conflicting views. (7)

The war took its toll on everyone; when it was over, German biology needed to be rebuilt virtually from scratch.

Finally, there is another metaphoric reference of nucleus and cytoplasm, surely the most conspicuous of all, and that is to be found in sexual reproduction. By tradition as well as by biological experience, at least until World War II nucleus and cytoplasm were also tropes for male and female.

15. It had originally been scheduled to meet in Russia, but the Russians canceled it amid the turmoil of the genetics purges instigated by Trofim Lysenko.

Until the emergence of bacterial genetics in the mid-1940s, all research in genetics and embryology, both in Europe and the United States, focused on organisms that pass through embryonic stages of development; for these organisms, a persistent asymmetry is evident in male and female contributions to fertilization. The female gamete, the egg, is vastly larger than the male gamete, the sperm. The difference is the cytoplasm, deriving from the maternal parent (a no-man's-land indeed); by contrast, the sperm cell is almost pure nucleus. It is thus hardly surprising to find that in the conventional discourse about nucleus and cytoplasm, *cytoplasm* is routinely taken to be synonymous with *egg*. Furthermore—by an all too familiar twist of logic—the nucleus was often taken as a stand-in for sperm.

Theodor Boveri argued for the need to recognize at least some function for the cytoplasm on the ground of "the absurdity of the idea that it would be possible to bring a sperm to develop by means of an artificial culture medium" (published posthumously in 1918 and translated in Baltzer 1967:83–84; see also Wilson [1896:262]). Thus many debates about the relative importance of nucleus and cytoplasm in inheritance inevitably reflect older debates about the relative importance (or activity) of maternal and paternal contributions to reproduction, where the overwhelming historical tendency has been to attribute activity and motive force to the male contribution while relegating the female contribution to the role of passive, facilitating environment. In Platonic terms, the egg represented the body and the nucleus the activating soul. (In a related vein, E. B. Wilson's remarks about Morgan's early passion for embryology may also be worth noting: "It is an open secret that even now he sometimes escapes from the austere heights where *Drosophila* has its home in order to indulge in the illicit pleasures of the egg and its development" (1932:82). I suggest that in these associations surely lies part of the background both for the force of the assumption of gene action and perhaps even its gradual fading away from the status of self-evident truth. More specifically, I suggest that such associations bear quite directly on the historic discounting of maternal effects.[16]

✤

Change, of course, did not come overnight. Although embryology was no longer a thriving research enterprise

16. Between the 1950s and the 1970s the term *maternal effects* disappeared from almost all indexes; for most authors the term *maternal inheritance* (in large part a direct consequence of the move of geneticists away from the study of metazoan organisms, especially *Drosophila*, and to the study of unicellular, asexually reproducing organisms) is replaced by *cytoplasmic inheritance*. Indeed, Ruth Sager, who studied inheritance in the single-celled protist *Chlamydomonas*, lists *maternal inheritance* in the index to her 1961 book with Francis Ryan, *Cell Heredity*, only to refer the reader to *uni-parental inheritance*. Unfortunately, such relabeling did little to protect her from the jokes of her colleagues, who derisively referred to her work as "Ruth's defense of the egg" (Sapp 1987:206).

after the war, the memory of that disciplinary struggle took time to abate. It also took time—roughly two decades—for German biology to rebuild. Last, it took the women's movement to change our ideas about gender and perhaps the hiatus of bacterial genetics (where no one had to think about male and female contributions) for these changes to creep into biology. By the time that the study of higher organisms began to reemerge in the 1970s, the entire world had changed, and so did the ways that seem natural to talk. Embryology was no longer a rival, Germany had become a friend, and gender equity was all the rage. There were of course other changes, which I have not talked about—most notable perhaps, the emergence of a discourse of feedback and of bodies as cyborgs, both associated with the extraordinary developments in systems analysis and computer science. And last but hardly least were the equally extraordinary developments internal to molecular biology, especially the techniques of recombinant DNA.

Concurrent with the changes in the way we talked—and thought—these developments soon effected dramatic changes in what could be done in the lab. During the last decade the world of technical feasibility has changed beyond recognition. These different kinds of changes—in how we talk and in what can be done in the lab—have worked in concert and in mutual reinforcement, the one creating the opportunity and the other the need to radically rethink the meaning of genetic control. But as broadly as I have drawn the map to try to account for this shift in biological discourse, it is still not broad enough. Other developments, both social and technical, also intervened. For example, we have yet to understand just how radically the advent of the computer has reconfigured our ways of thinking about bodies. At the very least, we must note how little resemblance the organism that molecular biology has rediscovered bears to the organism that the old embryologists sought to understand.

But my main point should be clear: acting in sync (as they always do), the social, cognitive, and technical histories of twentieth century biology have once again brought us to a dramatic and critical juncture. And if there is a moral to this story, it is this: lest we be too quick to congratulate ourselves for our new found enlightenment, we should remember that our predilections—grounded though they must be in our particular social and political realities—are all we have to guide us. Thus there is no guarantee that new doctrinaires will not seize the opportunity now before us; indeed, we have every reason to expect that they will—even to suspect that they already have. How else, after all, could science possibly proceed?

MOLECULES, MESSAGES, AND MEMORY:
LIFE AND THE SECOND LAW

And if mind itself requires a material habitat then it has in an atom an imperishable living home.

—A. E. Dolbear
"Life from a Physical Standpoint"

It is very surprising that laboratory scientists should still entertain this mirage, that it is the individual, the human subject . . . who is truly autonomous, and that, somewhere in him, be it in the pineal gland or elsewhere, there's a signalman, a little man within a man, who makes the apparatus tick. . . . It's like Archimedes—you give him his little point outside of the world, and he can move it. But this little point doesn't exist.

—Jacques Lacan, Seminars

The conception of genes as autonomous actors—endowed with the authority and the capacity to direct the future course of organismic development—dates to the mid twenties, but the particular twist added by the notion of the chromosome as "code-script" arrives only in the early forties. Credit for the introduction of this idea to biology belongs to Erwin Schroedinger, the physicist often cast as the father of quantum mechanics.

But Schroedinger was not merely a physicist; in his heart of hearts he was also a philosopher. And one of his major lifelong preoccupations was with the nature of heredity—as he saw it, a transmission from past to future oblivious of individual mortality, a genealogical memory immune to the ravages of time. For a physicist, the ravages of time were spelled out with unmistakable clarity in the second law of thermodynamics, a law that Sir Arthur Stanley Eddington held to be the highest of all the laws of nature. In Eddington's own words,

> The law that entropy always increases—the second law of thermodynamics—holds, I think, the supreme position among the laws of Nature. If someone points out to you that your pet theory of the universe is in dis-

agreement with [James Clerk] Maxwell's equations—then so much the worse for Maxwell's equations. If it is found to be contradicted by observation, well, these experimentalists do bungle things sometimes. But if your theory is found to be against the second law of thermodynamics I can give you no hope; there is nothing for it but to collapse in deepest humiliation. (1928:74)

Given Schroedinger's concerns, the problem was obvious and, given the status of the second law, it was serious: how could one account for the extraordinary stability of genetic memory in a world in which everything else was mere grist for the relentless forces of dissipation? How was one to reconcile this supreme law of nature with the equally undeniable fact of life?

In 1943, exiled in Dublin, Schroedinger tackled this problem in a set of lectures under the rather grandiose title "What Is Life?"—just the sort of metaphysical question upon which English-speaking scientists, at least, had by then looked so askance. But Schroedinger had an answer. Inspired by his recent encounter with an atomic model of gene structure proposed by N. W. Timoféeff-Ressovsky, K. C. Zimmer, and Max Delbrück (1935), and picking up on an argument he had sketched a decade earlier in Berlin, Schroedinger proposed that the solution to this apparent paradox, the peculiar power of living systems that enables them to resist the second law, which makes and keeps them "alive" in their individual lifetimes—and, more important, endows them with immortality through the mechanism of genetic memory—lies in the special structure of their chromosomes:

It is these chromosomes . . . that contain in some kind of code-script the entire pattern of the individual's future development and of its functioning in the mature state. Every complete set of chromosomes contains the full code. . . . In calling the structure of the chromosome fibres a code-script we mean that the all-penetrating mind, once conceived by Laplace, . . . could tell from their structure whether the egg would develop, under suitable conditions, into a black cock or into a speckled hen. . . .

But the term code-script is, of course, too narrow. The chromosome structures are at the same time instrumental in bringing about the development they foreshadow. They are law-code and executive power—or, to use another simile, they are architect's plan and builder's craft—in one. (1944:22–23)

My point in this essay will be to argue that this notion of code-script, inserted by Schroedinger into the chromosomes, brings Jacques Lacan's "little man within a man" into twentieth-century molecular biology, that it retrieves for contemporary biological discourse that Archimidean point with which we seem unable to dispense. But to make sense of Schroedinger's intervention, to better understand his notion of code-script—"law code and executive power . . . in one"—we must turn back in time to the nineteenth century and the introduction of the second law of thermodynamics in 1850 and especially to the debate this law sparked concerning the place of living systems in the laws of physics. Therefore I begin by briefly tracing the history of this debate from its inception in the middle of the nineteenth century.[1]

Demons Great and Small

The genesis of the second law was, of course, in the practical problem of increasing the efficiency of steam engines, but just because its thrust was to negate the possibility of perfectly efficient transformation of heat into motive power (in other words, without dissipation of energy), from the very start the question of how a living organism manages such effective conversion of the sun's heat into its own animate power lay close at hand. Indeed, William Thomson (later Lord Kelvin) derived the second law on the basis of a premise that explicitly excluded animate processes. As he wrote in his 1851 paper, "On the Dynamical Theory of Heat":

1. In this review, I lean heavily on the accounts of others, perhaps especially on Crosby Smith and Norton Wise's excellent work, *Energy and Empire* (1989).

The demonstration of the second proposition is founded on the following axiom:—

It is impossible by means of inanimate material agency, to derive mechanical effect from any portion of matter by cooling it below the temperature of the coldest of surrounding objects. (1911:179)

Reviewing the particular example of heat produced by electrical currents a year later, he offered two possibilities for this exemption of living organisms: either heat "in the animal frame acts so as to produce mechanical effect, from some of that heat"—with no fall of temperature, thereby violating the second law—or "the will of the animal can make these currents produce mech [sic] effect" directly, with heat as a mere side effect and no violation of the second law. "There is a great degree of probability in favour of the 2nd of these 2," he said during lectures he gave in 1852–1853. "If there be one or other of these hypotheses, the 2nd is nearly established" (quoted in Smith and Wise 1989:616). Thomson addresses the question of animal will even more explicitly in his 1852 essay entitled "On the Power of Animated Creatures over Matter" (1911):

The application of Carnot's principle, and of Joule's discoveries . . . is pointed out; according to which it appears nearly certain that, when an animal works against resisting force, there is not a *conversion of heat into external mechanical effect*, but the full thermal equivalent of the chemical forces is *never produced*; in other words, that the animal body does not act as a *thermo-dynamic engine*. . . .

Consciousness teaches every individual that they are, to some extent, subject to the direction of his will. It appears, therefore, that animated creatures have the power of immediately applying, to certain moving particles of matter within their bodies, forces by which the motions of these particles are directed to produce desired mechanical effects. (507–9)

But even if "the will of the animal" can exempt an individual living being from the inevitable forces of dissipation, life in general will not be so fortunate. In his 1852 paper, "On a Universal Tendency in Nature to the Dissipation of Mechanical Energy" (1911:511–14), Thomson discussed its cosmological implications, concluding that "within a finite period of time to come the earth must again be, unfit for the habitation of man" (514). Coining the term *thermal death,* Hermann Ludwig Ferdinand von Helmholtz reiterated this somber message two years later, warning that the universe is tending toward a state of thermal death in which the temperature will be both too low and too uniform to permit life to continue. Fifty years later Ernest Rutherford recapitulated: "Science offers no escape from the conclusion of

Kelvin and Helmholtz that the sun must ultimately grow cold and this earth become a dead planet moving through the intense cold of empty space" (quoted in Smith and Wise 1989:551).

On two grounds, then—first, on the individual, and, second, on the cosmological—the second law posed a threat to life. Gillian Beer has written eloquently of the acute anxieties in the Victorian age about "the death of the sun" and the particular challenge such a prospect posed both to popular confidence in the tendency toward at least human perfection and to Darwin's faith in progressive evolution on a grander scale. In 1876 Darwin recorded his own dismay in his *Autobiography*:

> Believing as I do that man in the distant future will be a far more perfect creature than he now is, it is an intolerable thought that he and all other sentient beings are doomed to complete annihilation after such long-continued slow progress. (153–54, quoted in Beer 1989:160)

Yet even though it failed to console Darwin, some sort of solace, both for the individual and the species, soon became available in a new form of mentalism—both more durable and more powerful than entropy. For the progression of individual life, this new escape from death was already hinted at by Thomson in his reference to "the will of the animal," and on the cosmic scale it was implicit in the focus of another physicist, James Clerk Maxwell, on the peculiar properties of molecules.

First, and more simply, the cosmic scale: between the promise of progressive development in Darwin's theories and the threat of inexorable decay and dissolution in the laws of thermodynamics, Maxwell foresaw a third possibility in the nature of molecular structure—neither of progress nor of decay but of stability. (Indeed, as I argue later, it is possible to trace a route from considerations of this third option directly to Schroedinger's speculations seventy years later about the specially durable molecular structure of chromosomes.) In his 1870 essay on "Molecules," Maxwell wrote,

> No theory of evolution can be formed to acct for the sim-
> ilarity of molecules, for evolution necessarily implies
> continuous change, and the molecule is incapable of
> growth or decay, of generation or destruction. . . .
>
> They continue this day as they were created, perfect
> in number and measure and weight, and from the inef-
> faceable characters impressed on them we may learn
> that those aspirations after accuracy in measurement,
> truth in statement, and justice in action, which we
> reckon among our noblest attributes as men, are ours
> because they are essential constituents of the image of
> Him who in the beginning created, not only the heaven

and the earth, but the materials of which heaven and earth consist. (quoted in Beer 1989:172–73)

As Beer concludes:

> The indivisibility of molecules has a mythic function. It becomes a way of sustaining the mind drifting towards dissolution with the cooling sun. The indivisible world is steadfast even when life-giving light dies out. (173)

But the mere indivisibility of molecules, however promising it may be for cosmic endurance, offers little solace for the individual life. For this, the force of entropy must be more directly countered. In a letter written to P. G. Tait in 1867, Maxwell offered a glimmer of a solution in his invention of a purely imaginary being who could reverse the natural tendency toward dissipation: "Conceive a finite being who knows the paths and velocities of all the molecules by simple inspection but who can do no work except open and close a hole in the diaphragm [between two vessels] by means of a slide without mass." If this finite being then permits only fast molecules to pass from the cooler vessel into the warmer one, blocking the slower molecules, the effect will be that "the hot system has got hotter and the cold colder and yet no work has been done, only the intelligence of a very observant and neat-fingered being has been employed" (reprinted in Knott 1911:213–14). With this phantasmatic conception enters one of the most extraordinary characters in the history of physics (perhaps especially in the relation of physics to biology), a being, "observant and neat-fingered," whom Thomson was soon to dub a Demon.

To be sure, Maxwell's Demon had precursors—most obviously in the Laplacean Intelligence but also in the rather less well known Being Darwin himself had introduced in his "Essay of 1844." There Darwin had written:

> Let us now suppose a Being with penetration sufficient to perceive differences in the outer and innermost organization quite imperceptible to man, and with forethought extending over future centuries to watch with unerring care and select for any object the offspring of an organism produced under the foregoing circumstances; I can see no conceivable reason why he should not form a new race . . . adapted to new ends. (quoted in Schweber 1982:322)

However, both Darwin's Being and Pierre-Simon Laplace's Intelligence were transfinite figures, manifest surrogates for God. True, Darwin's Being resembles Maxwell's Demon in one respect: it, too, acts as a selecting agent. But the crucial point is that *its* perception requires penetration from without; *it* selects from on high. Darwin's Being is, in short, his personification of nature writ large, indeed writ infinite. In contrast, what is distinctive to Maxwell's Being (perhaps what makes it a demon) is precisely its location inside the gas, on a level with the very molecules it must select; it is this "worldly finitude," indeed its minuteness. "Very small BUT lively" is how Maxwell put it in his notes (cited in Schweber 1982:7). Maxwell's Demon may be intelligent and endowed with free will, but equally critical (and distinctive) to its character (as Thomson and Maxwell both repeatedly emphasize) is its mechanical rather than preternatural nature. Ultimately, I will argue that for the successors of Maxwell's Demon—unlike its precursors, particularly as they figure in the problem of life—the most important point is its shift in reference from God to humanlike intelligence, a shift extended by Schroedinger ever more inward, until it becomes, in the mid-twentieth century, the ghost inside the molecular machine of life. But I am getting ahead of myself.

As yet, only the language of "intelligence" and an "observant and neat-fingered being" suggests a link between Maxwell's Demon and the status of living forms. To see the full strength of this link in the mentality of late nineteenth-century physics, we need to look further at the writings of these physicists, observing especially their free employment of a host of technological rather than theological images—images of railway pointsmen, steersmen, telegraph agents, and military strategists—in order to describe, alternatively, the properties of the Demon and those of mind, life, and free will.

Maxwell described his Demon as "a doorkeeper very intelligent and exceedingly quick, with microscopic eyes . . . like a pointsman on a railway with perfectly acting switches who sends the express along one line and the goods along another."[2] Similar language appears in his 1873 essay (in Campbell and Garnett 1882:434–44), not on the second law but on science and free will. Here it is the function of the Ego, which he likens not to the railway pointsman but to the steersman of a vessel, whose function is "not to produce, but to regulate and direct animal powers." He explains that

> in the course of this mortal life we more or less frequently find ourselves in a physical or moral watershed, where an imperceptible deviation is sufficient to determine into which of two valleys we shall descend. The doctrine of free will asserts that in some such cases the Ego alone is the determining cause. (quoted in Smith and Wise 1989:629)

Another Scottish physicist, Balfour Stewart, addressed the question of life yet more vividly in his textbook *The Conservation of Energy,* published the same year:

> Life is not a bully, who swaggers out into the open universe, upsetting the laws of energy in all directions, but rather a consummate strategist, who, sitting in his secret chamber, before his [telegraph] wires, directs the move-

2. Letter to J. W. Strutt (Lord Rayleigh), December 6, 1870, cited in Smith and Wise 1989:625.

ments of a great army. (1873, quoted in Smith and Wise 1989:629)[3]

The links that Thomson forges between the animating force of Life and the Demon are more explicit yet. In 1874, now as Lord Kelvin, he redefined Maxwell's Demon as "an intelligent being endowed with free will and fine enough tactile and perceptive organization to give him the faculty of observing and influencing individual molecules of matter" (1911, 5:12). And again, in 1879, as

> a creature of imagination having certain perfectly well-defined powers of action, purely mechanical in their character. . . . A being with no preternatural qualities and differs from real living animals only in extreme smallness and agility. . . . Just as a living animal does, he can store up limited quantities of energy, and reproduce them at will. (1911, 5:21–23)

Perhaps the principle point to note is that for (at least these) nineteenth-century physicists, Maxwell's Demon served a vitally dual function in their valiant struggle to maintain a conception simultaneously of free will and determinism, a

3. Stewart's earlier suggestion that the living being is "an organization of infinite delicacy, by means of which a principle in its essence distinct from matter, by impressing upon it an infinitely small amount of directive energy, may bring about perceptible results" echoes the properties of the Demon even more directly (1868, quoted in Smith and Wise 1989:628).

view of living beings as continuous with, yet distinct from, inanimate matter. As Kelvin tells us, the demon is "a being with no preternatural qualities and differs from living animals only in extreme smallness and agility" (1911, 5:21–23). Its escape from the second law of thermodynamics is effected by nothing more than minute intelligence, agility, and, perhaps most important, a frictionless slide. By the end of the nineteenth century, however, this strain began to prove too great to maintain. Kelvin, who had long been in some tension with Darwin over the latter's abandonment of intelligent and benevolent design, ultimately concluded that Life would have to be altogether exempt from the laws of physics. In 1892 Kelvin wrote,

> The influence of animal or vegetable life on matter is infinitely beyond the range of any scientific inquiry hitherto entered on. Its power of directing the motions of moving particles, in the demonstrated daily miracle of our human free will, and in the growth of generation after generation of plants from a single seed, are infinitely different from any possible result of the fortuitous concourse of atoms. (Smith and Wise 1989:644)

And in 1899,

> Mathematics and dynamics fail us when we contemplate the earth, fitted for life but lifeless, and try to imagine the commencement of life upon it. This certainly did not

take place by any action of chemistry, or electricity, or crystalline grouping of molecules under the influence of force, or by any possible kind of fortuitous concourse of atoms. We must pause, face to face with the mystery and miracle of living creatures. (612)

In the end, however, it was human agency rather than the commencement of life that provided Kelvin with the most powerful argument against a purely mechanical explanation of life and hence for the radical limitations of materialist doctrine. Just a year before his death in 1907, he wrote,

The perception of every one of the human race of his own individuality and free will seems to me to absolutely disprove all materialistic doctrines and to give us scientific ground for believing in the Creator of the Univ. in whom we live & move & have our being. (645)

In Kelvin's final view, even Maxwell's Demon would not suffice—at least not by itself—for beings such as ourselves.

Maxwell's Demon Lives On

Even in the face of Kelvin's misgivings, however, the Demon endured, albeit not without undergoing considerable metamorphosis. In a 1990 overview, Harvey Leff and Andrew Rex attribute three separate lives to Maxwell's Demon (at least until the present). They date the end of its first life—what I might call its life within Life (the life with which I am primarily concerned in this essay)—and the beginning of its second to 1929 when the Demon resurfaces inside a perpetual motion machine. The occasion is a key paper by Leo Szilard entitled "On the Decrease of Entropy in a Thermodynamics System by the Intervention of Intelligent Beings"—a paper that is retrospectively read as the origin of information theory, as anticipating the ultimate residence of the Demon in the computer. I might quibble with the date Leff and Rex give, but there is little question that the life of the Demon within Life—the attempt to locate the springs of vitality on a molecular level *inside* the organism—undergoes a noticeable decline in the years after Kelvin's death.

Two reasons, each far too large a subject to do more than indicate here, seem to me to more or less account for this decline. First is the progressive widening of the schism between mechanism and vitalism. By the time Hans Driesch laid out his argument for his principle of entelechy in the Gifford lectures in 1907 (published as *The Science and Philosophy of the Organism* a year later), the battle lines were already in place, and as the century wore on, they became ever more firmly delineated and hence more difficult to transgress. The life of the Demon within Life, inside the organism, could be found only on what, for most twentieth-century physicists, was the wrong side of the line. *"Es gibt seine Dämonen,"* Driesch cried, *"wir selbst sind sie"*

[Demons exist, we ourselves are them, quoted in Needham 1943:216].

The second reason bears on certain radical instabilities inherent in readings both of the Demon and of the second law. One locus of ongoing debate centered on the question of whether the Demon itself (or himself?) must perform work and hence produces entropy after all. (In the parlance of life the question becomes whether the organism must produce entropy in its ambient environment.) If the answer to this question in either form is yes, it can be said that the second law is at least globally obeyed, both by Maxwell's Demon and by living organisms. In a number of commentaries, the answer to this question is simply asserted as yes. But I know of no effort, before Szilard's 1929 paper, to actually perform the calculation. The import of Szilard's paper—and the reason for claiming it as (at least the first) death knell for Maxwell's Demon—was precisely to show that for a "biological system," the model of which was a piston operating in a one-dimensional gas, the Demon would in fact have to perform work in the process of measurement, thereby saving the second law.

Even so, some physicists continued to disagree (indeed, continue to disagree to this very day), both about the activities of the Demon and about Life. It is worth noting that all these debates and discussions (including Szilard's) were based not on the interpretation of observed phenomena but on the outcome of thought experiments—of scenarios existing only in the imagining. A particularly remarkable exchange between Sir James Jeans and the physical chemist F. G. Donnan attests to the confusion. In four rounds of increasingly technical letters published in *Nature* in 1934, Jeans and Donnan insist on opposite answers to the question of whether the total (global) entropy production of metabolic processes—the organism in interaction with its environment—need be positive (Jeans claims no, Donnan yes). The debate itself produces a great deal of heat but ends where it begins, with Jeans expressing the hope "that this discussion will end soon," and Donnan attempting to recruit allies sufficiently numerous to counteract Jeans's otherwise indisputable authority. Donnan's last words are

> Though we have discussed the matter with several authorities, including one of the most eminent theoretical physicists of the world, we have failed to find anyone, other than Sir James Jeans, who disagrees with us. We are content, therefore, to leave the matter in dispute to the judgment of the readers of *Nature*. (August 18, 1934:255)

A second locus of at least tacit disagreement bore on the local accumulation of order in the living system itself and on whether such a local accumulation would be in accord with the second law. In most discussions that occurred in the first three decades of the century, the two loci of disagreement were deployed interchangeably, as if recognizing only a single question: for those inclined toward vitalism, the crucial question was the second, the local accumulation of order, whereas for those committed to mechanism, the only relevant question was the first, the global production of entropy. To provide a flavor of how this worked, I quote from a typical exchange, published in 1914. In support of vitalism, James Johnstone writes,

> Life, when we regard it from the point of view of energetics, appears as a tendency which is opposed to that which we see to be characteristic of inorganic processes.
> . . . The effect of the movement which we call inorganic is toward the abolition of diversities, while that which we call life is toward the maintenance of diversities. (314)

In opposition, Ralph S. Lillie retorts that

> the evidence that the second law of thermodynamics does not always apply to life-processes is certainly inadequate. . . . There seems in fact to be a fundamental misconception in this part of the author's argument. He holds that life may play the part of the Maxwellian demon under appropriate circumstances. . . . [But] even Maxwell's demon has to work a partition which resists the impact of the faster molecules. (1914:842)

And so it went, with vitalists insisting on a radical opposition between living processes and the second law and those who called themselves mechanists denying the existence of any such problem. A small group (Joseph Needham [a biologist], J. H. Woodger [philosopher], and David Watson [physicist]) valiantly attempted to shore up a third option, arguing that the crucial issue was not order but organization. They thought of themselves as neomechanists, seeking to establish the autonomy of biological questions from physics without, however, claiming any violation of the second (or any other) law. As Watson put it in 1931,

> The possibility that the second law of thermodynamics may not apply to living systems rests mainly on the idea that an organism . . . may function in a manner similar to Maxwell's Demon. . . . [That is,] the system in its later states is more highly organized. . . .
>
> [Although] It is generally believed that this mechanism cannot lead to a reversal of the second law of thermodynamics in living matter, . . . it appears that one of the outstanding characteristics of life shows certain similarities to the above [to the Demon]. . . . Life modifies the "organization" of matter. (148–49)

However, he quickly adds:

> I wish to avoid the suggestion that these considerations involve a possible interference with the second law as ordinarily understood, or that the Demon as a quasi-

intelligent agency is necessary for the creation of biological organization. (149)

It is difficult to know what to make of or even to charac-
terize the tenor of these arguments—elusive at best and certainly a far cry from the ideal I was taught in graduate
school. By such standards, none of the questions I have
mentioned was ever in fact resolved. Given the lack of con-
sensus about how entropy was to be measured or even
defined in a nonequilibrium system, one might say in retrospect that they could not have been resolved. Indeed, only
in the late twenties and early thirties did explicit distinctions
between equilibrium and nonequilibrium (or between
closed and open) systems—and the contemporaneous
characterization of living organisms as open, nonequilibrium systems—begin to appear in the literature.

However, one problem did seem to be settled, if by
repeated fiat rather than by analysis. As I've already suggested, if one redraws the boundary of the living system,
not at the outer skin of the organism but at the outer
perimeter of a closed thermodynamics system large
enough to encompass the energetic substrates required for
respiration and metabolism, the problem might be said
(and indeed was said) to vanish. Such a redefinition of the
system served simultaneously to restore harmony between
biology and physics and to evade the ever more discredit-
ing charge of vitalism. (Only a few decades later, Francis

Crick would assign the vestiges of vitalism to the "lunatic fringe" [1966:99].) For the record, however, I must stress that the particular issue of the thermodynamic properties of that problematic subsystem, the individual organism, was never settled, neither by evidence, nor by argument, nor even by fiat. By the eve of World War II, geneticists had largely shelved discussions of the special status of living organisms, particularly in relation to the second law, perhaps in the hope that, along with Jeans, it would just go away.

Schroedinger's Resolution

Schroedinger, however, had not forgotten. With his well-honed credentials—no one would mistake the father of quantum mechanics for a vitalist—he confronted the question of Life and the second law head on. He did not even mention Maxwell's Demon but simply asked, "How can the events *in space and time* which take place within the spatial boundary of a living organism be accounted for by physics and chemistry?" (1944:3). Or, what, in the language of physics and chemistry, is life? "What is the characteristic feature of life? When is a piece of matter said to be alive?" (74).

His answer was simple and pointed directly to the problem of the second law:

When it goes on "doing something," moving, exchanging material with its environment, and so forth, and that for a much longer period than we would expect an inanimate piece of matter to "keep going" under similar circumstances. When a system that is not alive is isolated or placed in a uniform environment, all motion usually comes to a standstill very soon as a result of various kinds of friction. . . . After that the whole system fades away into a dead, inert lump of matter. . . . The physicist calls this the state of thermodynamic equilibrium, or of "maximum entropy." (74)

Thus the key issue for Schroedinger was "How does the living organism avoid decay? The obvious answer is: By eating, drinking, breathing and (in the case of plants) assimilating" (75).

But surely, neither mere material nor caloric intake will suffice to keep the organism alive; he wrote, "That the exchange of material should be the essential thing is absurd" (76). Rather, it is a "precious something contained in our food which keeps us from death" (76). A living organism "can only keep aloof from [the dangerous state of maximum entropy], i.e. alive, by continually drawing from its environment negative entropy. . . . What an organism feeds upon is negative entropy" (76). "How," Schroedinger went on to ask, "should we express in terms of the statistical theory the marvelous faculty of a living organism, by which it delays the decay into thermodynamic equilibrium [death]?" (78). The father of quantum mechanics found the answer: "The device by which [it] maintains itself . . . really consists in continually sucking orderliness from its environment" (79).

Let me underscore some of Schroedinger's words: The marvelous faculty of a living organism—that which guarantees its existence—is a *device*. What kind of device *really*, as he put it? This he did not yet say—only that that it "really consists in continually sucking orderliness from its environment." Is it possible that the silent ego behind the cogito, and hence behind the sum, is no more (or less) than an infant sucking life from the breast of its world? Not quite. *Really*, we have been instructed, this is not an infant homunculus; it is a *device*.

However, closer scrutiny seems to reveal it is not an infant at all but an old man; perhaps it is Mr. Schroedinger himself. Certainly, there is an oddity to this text—one that might have gone unnoticed, were it not for the surfacing of the figure of the infant. That oddity has to do with where Schroedinger (in a curiously personalized echo of Victorian anxieties about the "death of the sun") located the center of the paradox between life and the second law— not in the formation of life, not in embryogenesis and its conspicuously associated increase of order, but in the end of life and the impending breakdown of order that comes with death. The "characteristic feature of life" is its resistance to deterioration—not its capacity for reproduction, growth, or development but its ability to "keep going" for so very long. For Schroedinger, the miracle of life was not birth but the staving off of death; it was the "organism's astonishing gift of concentrating a 'stream of order' on itself . . . —of 'drinking orderliness' from a suitable environment" (82). This—and the parallel mystery of genetic immortality, the ability to "keep going" through progeneration—is at once the miracle of life, and the enigma Schroedinger had to solve.[4]

As he reminds us, "When a system that is not alive is isolated or placed in a uniform environment, all motion usually comes to a standstill"(74). The question is how organisms manage to avoid the inexorable fate of their inanimate counterparts. How does a living being, similarly isolated, perhaps deserted, "keep [on] going"? "What is the precious something . . . which keeps us from death?" (76).

For anyone but a physicist, we might be tempted to insert

4. For Schroedinger the thermodynamic problems of individual life and the life of the species were interchangeable. Just as he displaced the problem of increasing order in embryogenesis onto that of "staying alive," so too did he displace the thermodynamic problem of the commencement of life on the planet onto that of the genetic continuity of the species.

a bit of context. The year, after all, was 1943. Europe was in flames, but Schroedinger, fifty-six years old, kept on going. Four years into what he called his "long exile" from Germany, he at least was safe. In 1939 he had (narrowly) escaped from Nazi rule and the war, finding refuge in and on the island nation of Ireland—beautiful, green, and politically neutral. Many years later, he reminisced that

I would never have gotten to know this remote and beautiful island otherwise. Nowhere else could we have lived through the Nazi war so untouched by problems that it is almost shameful. I can't imagine spending seventeen years in Graz "treading water," with or without the Nazis, with or without the war. Sometimes we would quietly say amongst ourselves: "Wir danken's unserem Fuhrer." ([1960] 1992:183)

To be sure, Schroedinger had been preoccupied for some time by the question of the special laws (the *Eigengesetzlichkeit*) of living matter and their relation to physics; he had been in correspondence on the subject with the British physical chemist F. G. Donnan since the mid-1930s.[5] But his earlier musings lacked the preoccupation with survival that

5. I am grateful to the librarian of manuscripts and rare books at The Library, University College, London, for sending me photostats of the relevant correspondence between Schroedinger and Donnan.

so marked his 1943 lectures.[6] Now, surrounded by death as well as by the sea, perhaps even "treading water," and certainly "going on," he recalled a quote from Spinoza: "There is nothing over which a free man ponders less than death; his wisdom is to meditate not on death but on life" (1944:2). And in his meditation, Schroedinger discovered that the question of how we stay alive is "easily answered." The "precious something" that "keeps us from death" is negative entropy, which is in fact "something very positive":

> The essential thing in metabolism is that the organism succeeds in freeing itself from all the entropy it cannot help producing while alive. (1944:76)

Yet, for all of Schroedinger's "going on," the question still begs: How *does* the organism actually accomplish this

6. In one of his letters to Donnan, written from a temporary domicile in Belgium on September 7, 1939, while he was en route from Germany, Schroedinger focused his attention on the particular problem the second law poses, not to survival but to the origin of life, praising Donnan for his moderate position on this problem. Halfway into the letter, he recalls where and when he is writing, and he stops:

Am I awake or am I dreaming? Are these the most relevant things to discuss four days after the opening of what may turn out to be the last fight of civilised mankind against the inferno? After the beginning of what may eventually be the twilight of civilisation or its true dawn?

I have a strange feeling. I am sitting here idle in a beautiful sea side resort, which is emptying quickly, for everybody seems to have something important to do, whereas I am entering the fourth month of involuntary vacation.

feat? By what process does it free itself from entropy, always drawing, sucking, drinking, concentrating order from its environment? What, in short, *is* this device, and how does it work? Slipping without comment into the paradox of genetic continuity, he wrote, "An organism's astonishing gift of concentrating a 'stream of order' on itself and thus escaping the decay into atomic chaos . . . seems to be connected with the presence of the 'aperiodic solids,' the chromosome molecules" (1944:82). The device for staving off death—in one breath, of both the individual and the species—is to be found in the gene and, more specifically, in its "peculiar arrangements of molecules to which the ordinary laws of physics do not apply" (letter to Donnan, October 26, 1942).

Traces of Maxwell's Demon

As I've said, Schroedinger made no reference to Maxwell's Demon. But I am not certain that he managed to solve the problem of Life and the second law without the help of something very like Maxwell's Demon. The deus ex machina in Schroedinger's solution, I suggest, returns hidden in the structure of "the most essential part of a living cell," the locus of genetic memory, the chromosome or gene. Reminiscent of Maxwell's appeal to molecular stability as a means of keeping mind and matter steadfast in a dying world, Schroedinger wrote,

To reconcile the high durability of the hereditary substance with its minute size, we had to evade the tendency to disorder by "inventing the molecule," in fact,
an unusually large molecule which has to be a master-
piece of highly differentiated order, safeguarded by the
conjuring rod of quantum theory. (1944:73)

(Maxwell, it would appear, was just whistling in the dark—
Schroedinger reminded us repeatedly that until the advent
of quantum theory, physics had no way to account for the
stability of molecules.)

This molecule is remarkable indeed. Better than a
homunculus (because it is safeguarded by quantum theory),
it contains a miniature code that is "in one-to-one correspondence with a highly complicated and specified plan of
development" (1944:72). That is, it need not actually *be* the
body, even in miniature; it need only contain a *plan* for executing the development of that body.

But what makes this molecule truly remarkable is not in
fact its complexity, written into its aperiodic (codelike)
structure. As Schroedinger put it, this molecule must also
"somehow contain the means of putting [this plan] into
operation." The device that accounts for an "organism's
astonishing gift of concentrating a 'stream of' order on itself
. . . —of 'drinking orderliness' from [its] environment" must
not simply represent a plan for the "regular and lawful . . .
unfolding of events in the life cycle"; it must also be instru-

mental in the carrying out of that plan. It must contain a mechanism that guides, controls, and executes all the necessary events. It must, in short, be "law-code and executive power" in one (1944:82).

There remains, of course, the problem of multicellularity. That is, the bodies of most actual organisms are constituted of a great many cells, each containing a copy of the same codescript (or device). How, then, are these multiple devices coordinated? Schroedinger rephrased this problem as one of communication, and he offered two analogies—one military, the other political. Early on, he remarked (in the only reference to current events that appeared in the entire work):

> Some time ago we were told in the newspapers that in his African campaign General Montgomery made a point of having every single soldier of his army meticulously informed of all his designs. If that is true . . . it provides an excellent analogy to our case, in which the corresponding fact certainly is literally true. (1944:24)

In his final pages, he turns to political analogy to account for the same effect:

> Since we know the power this tiny central office has in the isolated cell, do they not resemble stations of local government dispersed through the body, communicating with each other with great ease, thanks to the code that is common to all of them? (84)

The image is clear: in every cell is to be found a represen-
tative (be it a soldier or local government station) of the
same code-script, the same "law-code and executive
power," empowered by its progenitor (general or central
government office) to guide, control, and execute the construction and maintenance of the body under its immediate
charge. Coordination of these agents is guaranteed by having all soldiers carry identical manuals—in fact, there is no
need for communication in the ordinary sense of the word;
rather, *communication* means nothing more here than repetition. And, somewhat mysteriously, the organism—the
collective body—with all its parts in the appropriate size
and shape and with all the necessary functions in operation—emerges from this repetition as a natural by-product.

✻

My first lecture was devoted to the biological discourse of
gene action. Schroedinger added to this discourse several
metaphors, already familiar from nineteenth-century discussions of Maxwell's Demon: he figured the uncanny order of
the body as a product of an intelligence system—such as the
army or the state—indeed, a system also quite akin to the
systems of "command, control, and communication" that
Norbert Wiener and his colleagues were just at that time
beginning to develop. Schroedinger invoked neither railway
pointsmen, nor intelligent doorkeepers, nor miniature
"observant and neat-fingered beings"; yet the devices he

invoked to perform the function of "concentrating order," the agents that must perform the work of generals and executives, are not far removed. They must be very small, of molecular size; indeed, these intelligent agents *are* molecules.

In the epilogue to *What Is Life?*—an epilogue so embarrassing to his contemporaries that his friends attempted to persuade him against publishing it, and most readers since have simply chosen to ignore it—Schroedinger's meaning becomes considerably clearer. Here he returned to the more traditional and transcendent issue of mind, to the problem of "Determinacy and Free Will":

> According to the evidence put forward in the preceding pages the space-time events in the body of a living being which correspond to the activity of its mind . . . are . . . if not strictly deterministic at any rate statistic-deterministic. (1944:92)

But our undeniable introspective experience tells us we are free. How are we to reconcile such a deterministic view with the fact of free will? *"What is this 'I'?"* Schroedinger wrote (1944:96). For a physicist, contradiction is the ultimate prohibition. Accordingly, he suggested:

> So let us see whether we cannot draw the correct, non-contradictory conclusion from the following two premises:

(i) My body functions as a pure mechanism according to the Laws of Nature.

(ii) Yet I know, by incontrovertible direct experience, that I am directing its motions. . . .

The only inference from these two facts is, I think, that I—I in the widest meaning of the word, that is to say, every conscious mind that has ever said or felt 'I'—am the person, if any, who controls the "motion of the atoms" according to the Laws of Nature. . . .

In Christian terminology to say: "Hence I am God Almighty" sounds both blasphemous and lunatic. But please disregard these connotations for the moment and consider whether the above inference is not the closest a biologist can get to proving God and immortality at one stroke. (93)

Clearly, Lacan's observation, with which I began this essay, could not have been more to the point. Schroedinger's subject—his Demon, or Laplacean Intelligence, his very Self—is not in the pineal gland but elsewhere—in the molecular structure of the gene. It is there that we will find the "signalman, a little man within a man, who makes the apparatus tick." But perhaps I should have prepared my readers, as Schroedinger did his (although not the audience for his lectures), with some of his own chapter headings. Before chapter 1 appears nothing other than the quotation "Cogito ergo sum." And introducing chapters

2 and 3 are two equally appropriate quotes, now from Goethe:

> Chapter 2: Being is eternal; for laws there are to conserve the treasures of life on which the Universe draws for beauty.
>
> Chapter 3: And what in fluctuating appearance hovers, Ye shall fix by lasting thoughts.

Thought, then (or again), is the thing. But of course Schroedinger could not for long fix that supremely modernist subject within the boundaries of life. Other events—most notable the computer—were even then beginning to take over the management of his notion of a genetic code-script. Indeed, by the time he wrote, Maxwell's Demon—and, with it, thought itself—had already escaped the confines of Self. Over the years that followed, Thought and Life both have been thoroughly dispersed on the winds of information. As we well know, the computer is now the thing.

3

THE BODY OF A NEW MACHINE: SITUATING THE ORGANISM BETWEEN TELEGRAPHS AND COMPUTERS

> *But, as these primary constituents [the genetic determinants] are quite different from the parts themselves [the adult organs], they would require to vary in quite a different way from that in which the finished parts had varied: which is very like supposing that an English telegram to China is there received in the Chinese language.*
>
> —August Weismann in John Maynard Smith
> *"Weismann and Modern Biology"*

> *A gene, we may say, is a message, which can survive the death of the individual and can thus be received repeatedly by several organisms of different generations.*
>
> —H. Kalmus
> *"A Cybernetic Aspect of Genetics"*

✲

Genes and messages have a long association in biology, dating at least to Weismann. But through most of this history, even with the dramatic concreteness that molecular biology lent to this association, the image dominating most thinking about messages was drawn from the nineteenth-century technology of the telegraph. In the mid twentieth century, a new technology, the computer, arrived to displace the telegraph. With that displacement, the meanings of many terms—of *message,* of *information,* of *organization,* indeed, of the *organism*—have, over the last few decades, been transformed. Here I want to explore the computer's impact on biological representation of the organism in two disciplines, molecular biology and developmental biology.

Molecular biology began with a simple strategy, borrowed from a long tradition of success in physics: it sought to reduce its world, to find the essence of life in organisms so rudimentary and so simple as to be immune from the mystifying and recalcitrant chaos of higher complex organisms. This strategy, as we know, led molecular biologists to the study of life in test tubes and petri dishes populated by bacterial cultures of *E. coli* and their viral parasites known as *bacteriophage*—forms of life that seemed simple enough to preserve the linearity of simple codes and telegraphlike messages. Meanwhile, outside their labs, life—the life of political, economic, and military exchange—was daily growing more chaotic and vastly more complex. Indeed, the computer, along with systems analysis more generally, had already been developed in the effort to cope with this complexity. The *computer,* as we now understand the term, came into being during the war years, because those people (mostly women) who had by tradition been responsible for data processing and computation—the original meaning of the word *computer*—could no longer handle the masses of data required to coordinate wartime military operations.[1] Similarly, with telephone wires criss-crossing the globe, Bell Laboratories needed ways to maximize the efficiency and reliability of transmission, hence the quantification of information and the emergence of information theory, with *information* defined as the mathematical converse of entropy.[2] Jacques Lacan put it well:

1. An image frequently evoked to depict the early days of the computer revolution is a room full of women calculating artillery firing tables at Aberdeen Proving Ground in Maryland; it was at this site, and for this task, that one of the first differential analyzers was used.

2. Statistical measures of a quantity called *information* (and understood in terms of the transmission of meaning-independent signals) were introduced in the late twenties and early thirties (Hartley 1928; Lewis 1930), but it is to C. E. Shannon (1948) that credit is usually given for the beginnings of information theory as such.

The Bell Telephone Company needed to economize, that is to say, to pass the greatest possible number of communications down one wire. In a country as vast as the US, it is very important to save on a few wires, and to get the inanities which generally travel by this kind of transmission apparatus to pass down the smallest possible number of wires. That is where the quantification of communication started. . . . It had nothing to do with knowing whether what people tell each other makes sense. . . . It is a matter of knowing what are the most economical conditions which enable one to transmit the words people recognize. No one cares about the meaning. . . . The quantity of information then began to be codified. This doesn't mean that fundamental things happen between human beings. It concerns what goes down the wires, and what can be measured. . . . It is the first time that confusion as such—this tendency there is in communication to cease being a communication, that is to say, of no longer communicating anything at all—appears as a fundamental concept. That makes for one more symbol. (1988:82–83)

In the war years, communication was perforce in the service of control. Norbert Wiener made explicit the conjunction between communication and control. His 1943 paper with Arturo Rosenblueth and Julian Bigelow on active, purposive behavior in machines, equating teleology with negative feedback—the paper often credited with founding cybernetics—followed directly on his wartime work on self-guiding, goal-seeking gunnery. His book, *Cybernetics, or Control and Communication in the Animal and the Machine,* appeared in 1948, following the 1946 interdisciplinary Macy Conference that he and John Von Neumann led on "Feedback Mechanisms and Circular Causal Systems in Biology and the Social Sciences."

Systems engineering (along with what Tom Hughes calls a systems engineering style of management [1993]) provides my last example. This mode of analysis was developed to handle the overwhelmingly large and heterogeneous manpower systems both of modern corporations and, even more, of postwar military defense strategies. The air force's Atlas/Titan intercontinental ballistic missile project of the 1950s "involved 18,000 scientists, engineers, and technical experts in universities and industry; 70,000 people from office to factory floor in 22 industries; . . . 17 associated contractors and 200 subcontractors with 200,000 suppliers; and about 500 military officers with technical expertise" (Hughes 1993:12).

These disciplines—information theory, cybernetics, systems analysis, operations research, and computer science—might be thought of as loosely linked endeavors sharing a common task (the analysis of complex systems), a conceptual vocabulary for dealing with that task (*feedback* and *communication*—circular causality), and a mode of representation (of complex systems as interacting networks or circuits). Although undoubtedly different in important respects (and surely warranting separate examination for these differences), here I am interested primarily in their commonalities. To emphasize these, I will refer to these disciplines by the single term *cyberscience.*

Cyberscience

Cyberscience, then, was developed to deal with the messy complexity of the postmodern world, over the very same period of time in which molecular biology was crafting its techniques for analyzing the simplest strata of life. The one repudiated conventional wisdom about the analytic value of simplicity, whereas the other embraced it; the one celebrated complexity, whereas the other disdained it. Small wonder, then, that these two disciplinary structures diverged.

We might even say that they inhabited different worlds. For cyberscientists, Life (especially corporate life, electronic life, and military life—the modes of life from which these efforts emerged and on which they were focused) had become far too unwieldy to be managed by mere doing, by direct action, or even through the delegation of "doing" to an army of underlings kept in step by executive order.[3] The problems of the mid twentieth century were clearly, as Warren Weaver put it in 1948, ones of "organized complexity":

> These are all problems with a *sizeable number of factors which are interrelated into an organic whole* [emphasis in original]. . . .
>
> These new problems, and the future of the world depends on many of them, requires science to make a third great advance, an advance that must be even greater than the nineteenth-century conquest of problems of simplicity or the twentieth-century victory over problems of disorganized complexity. Science must, over the next 50 years, learn to deal with these problems of organized complexity.[4] (538–39)

3. Other authors have stressed the role of cybernetics in extending the regime of wartime power, of command-control-communication, to the civilian domain (see Paul Edwards, in press; Peter Galison, in press; and Donna Haraway 1981). By contrast, my focus here is on the development of cyberscience *in response to the increasing impracticability of conventional power regimes*—which of course is not to absolve cyberscience of its political and military context but rather to resist simultaneously a unidimensional conception of power and a unidimensional reading of technoscience.

4. Warren Weaver's preoccupation with the "new" problems of "organized complexity," and his advocacy of the need for correspondingly new scientific tools for dealing with such problems, is particularly instructive. An early advocate of the use of physics in the creation of a new "science of man," and coiner of the term *molecular biology,* Weaver showed remarkably little interest in the actual molecular revolution in biology that occurred in the 1940s and 1950s. By then, his interests had shifted. As a leader of the gun control section of the U.S. Office for Scientific Research and Development during World War II, Weaver was responsible for directing Norbert Wiener's attention to the problem of tracking, which in turn led directly to Wiener's own work on purposive machines.

Fortunately, he goes on to observe,

> Out of the wickedness of war have come two new developments that may well be of major importance in helping science to solve these complex twentieth-century problems.
>
> The first piece of evidence is the wartime development of new types of electronic computing devices. . . . The second of the wartime advances is the "mixed-team" approach of operations analysis. . . . [Both] are familiar to those who were concerned with the application of mathematical methods to military affairs. (540)

On the microscopic scale, however, life still seemed simple. Or so it appeared to molecular biologists, who, in their own relatively cloistered settings (insulated equally from problems of military coordination and from the "organized complexity" of multicellular organisms), retained the confidence that their microscopic forms of life, at least, could be managed by "gene action," by master molecules exerting unidirectional control, translating their miniature linear codes into the four dimensional structures (spatial and temporal) of living organisms.

A great deal has been written by historians of modern biology about the influence of physics and physicists on the development of molecular biology and especially on biological attempts to model the organism on traditional (clockwork) machines. But that influence is only part of the story. To understand the changes that have taken place since World War II in our understanding of what an organism is, we need to trace a far more complex web of interactions.

To be sure, the traffic from physics to molecular biology in the postwar period was both extensive and significant—in philosophy, in techniques, in legitimacy, and in actual bodies (one thinks, for example, of Francis Crick, Maurice Wilkins, Gunther Stent, Seymour Benzer, and perhaps especially of Leo Szilard and Max Delbrück). Equally without question, most of this traffic provided strong support for the reductionist commitments of molecular biology (the use molecular biologists made of Schroedinger's intervention provides a key case in point [see, for example, Yoxen 1979, Keller 1990]). But there was also a kind of reverse traffic between biology and the physical sciences of immense conceptual (or ideological) import to the postmodern age. We can see that reverse traffic in all the disciplines of cyberscience I have been discussing. While molecular biologists (many of whom had, as I say, come from physics) were struggling to build a new biology that would be clearly distinct from (and even in opposition to) an older, organismic biology, as they sought to rid their descriptions of living organisms of traditionally vital (or vitalistic) preoccupation with *function*—especially to expunge from their language such conspicuously teleological notions as *purpose, organization,* and *harmony*—a number of other physicists and engineers (especially at such places as the Massachusetts Institute of Technology) were actively importing those older preoccupations to the language of cyberscience. These other physicists and engineers were leaning heavily on the very images, language, and even conceptual models that premolecular (organicist) biological discourse had championed, just to support the development of new paradigms of circular feedback for cybernetics and systems theory.

There were also other kinds of traffic between physical science and biology, one of which especially warrants noting. I am thinking of work in which notions of networks and organizational complexity, borrowed from cyberscience by theoretically minded biologists, were reimported to biology in an effort to revitalize an older organicist conception of development. Such efforts were of interest not to the burgeoning field of molecular biology but to the then-quiescent field of developmental biology (which some still called embryology). Because these other streams of traffic have received so little attention from historians of biology, I will briefly describe them both here.

Systems as Organisms

Just three years before Watson and Crick, a "Progress Report of the Air Defense Systems Engineering Committee," dated May 1, 1950, and written for the scientific advisory board to the chief of staff of the U.S. Air Force, provides the first definition of *system* in modern technical usage; it also may be thought of as providing an early prototypic definition of a *cyborg*. It begins:

> The Air Defense System [ADS] has points in common with many of [Webster's] different kinds of systems. But it is also a member of a particular category of systems: the category of organisms. This word, still according to Webster, means "a structure composed of distinct parts so constituted that the functioning of the parts and their relation to one another is governed by their relation to the whole." The stress is not only on pattern and arrangement, but on that also as determined by function, an attribute desired in the Air Defense System.
>
> The Air Defense System then, is an organism. . . . What then are organisms? They are of three kinds: animate organisms which comprise animals and groups of animals, including men; partly animate organisms which involve animals together with inanimate devices such as is the ADS; and inanimate organisms such as vending machines. All these organisms possess in common: sensory components, communication facilities, data analyzing devices, centers of judgement, directors of action, and effectors, or executing agencies.

Organisms also have the power of development or growth. . . . Moreover, they require to be supplied with material. . . . Nearly all organisms can sense not only the outside world, but also their own activities. . . . It is the function of an organism to interact with and alter the activities of other organisms, generally to achieve some defined purpose. ("Progress Report" 1950)[5]

For these writers, the central features of a system were function, coordination, interdependence, and purpose, all represented par excellence by the biological organism. And circular feedback was the way to provide a model of these features.

The affinity between systems and organisms, and the circular feedback so central to both, was of course equally key to the development of cybernetics. As I've already indicated, Wiener's initial preoccupation was with how to design purposeful, self-steering, target-seeking devices, to build into machines the very capacity for active, purposive behavior that one saw in biological function. Looking to an older biological tradition—to physiological work on homeostatic mechanisms and to hormonal systems of "reciprocal influences"—his answer, too, rested on the principle of circular feedback. Notions of systemic, interactive behavior based on circular feedback were common coinage in the discussions among the participants of the Macy conferences that

5. Courtesy of Tom Hughes.

Wiener helped organize in the late 1940s (see Heims 1991). Indeed, the organizers of these conferences sought to reflect the same principle of dynamic interaction in the form of the intellectual exchange they encouraged; accordingly, they made a deliberate effort to include workers from all possibly relevant disciplines, including genetics. They invited Delbrück to represent genetics. Unfortunately, he attended only a single meeting, in 1948. In a letter he subsequently sent to Steve Heims (1973) in response to Heims's query about his participation, Delbrück explained:

> You understate if you guess that the broad interdisciplinary approach made the discussion too diffuse for my taste. It was vacuous in the extreme and positively inane. Genetics did not, and at that time could not, enter into it at all. Also I was not then, and have not later, been much interested in the areas of information processing.[6]

Geneticists were on a different track. They had based their hopes not on the harnessing of complex systems but on what has been for natural scientists the more traditional paradigm of control—on the epistemological and technological benefits of reductio ad simplicitatum. The power of molecular biology lay in identifying the simplest unit of analysis, on the construction of *E. coli* and bacteriophage as model organisms for the study of genetics and development (recall Jacques Monod's dictum,

6. Courtesy of Steve J. Heims.

"What's true for *E. coli* is true for the elephant" [cited in Judson 1979:613]). *E. coli* is small, simple, and, above all, culturally homogeneous (by which I mean that all cells in a bacterial culture are identical). It is by its very nature insulated from the heterogeneity that is so central to the organization of higher organisms—the problem of differentiation and development. Finally, and most important of all, geneticists were committed to finding the source of gene action.

In 1958 Francis Crick formulated what is surely the key principle of molecular biology: the Central Dogma. The crucial point of the central dogma is its insistence on unidirectional causality, its repudiation of the possibility of a substantive influence on genes, either from their external or from their intra- or intercellular environment. Instead of circular feedback, it promised a linear structure of causal influence, from the central office of DNA to the outlying subsidiaries of the protein factory. To this end, they appropriated the cybernetic term *information,* but used it in its colloquial rather than in its technical sense.[7] Crick writes, "Once 'information' has passed into protein *it cannot get out again*" (1958:153).

7. Shannon defined information as the negative of the quantity formally identical to thermodynamic entropy. So defined, information becomes explicitly independent of meaning or function. Molecular biologists may have leaned on the technical usage of information for legitimacy, but they required its colloquial usage to give the term meaning relevant to their own context (see footnote 13). Whatever payoff they acquired from the term was clearly not technical.

Contested Uses of Information

MOLECULAR BIOLOGY

With Watson and Crick's invocation of "genetical information" residing in the nucleic acid sequences of DNA, some notion of information (however metaphorical) assumed a centrality to molecular biology that almost rivaled that of the more technical definition of information in cybernetics. But there were at least two crucial differences. The first, already noted in chapter 1, involves the gulf between *genetical information* and Shannon's term; the second bears more directly on the meaning of *instruction.*

As Lacan observed, the quantitative measure of communication introduced by Shannon "had nothing to do with . . . whether what people tell each other makes sense. . . . No one cares about the meaning" (1988:82). Not so for molecular biologists, not even for Watson and Crick. If "genetical information" is to have anything to do with life, it must involve meaning. A point mutation in the genetic code, a change in a single base pair of DNA, would make no difference to Shannon's measure of information, but for an organism, it would almost certainly make just the difference that matters—the difference between life and death.[8]

With this shift in the meaning of *information* we can begin to better understand the conflation between *informa-*

8. André Lwoff made this point explicitly in his 1960 Compton lectures (1968:93–94; see also footnote 13).

tion and *instruction* that became, after Watson and Crick, so conspicuous and that proved so peculiarly productive in the literature of molecular biology. If the genetic code is a message, it is a very particular kind of message: it is an order.[9]

Cast in the imperative, it says, "Make an enzyme!" As Schroedinger so aptly observed, this is no ordinary code: it is "law-code and executive power in one" (1944:82). Indeed, its capacity to issue orders is precisely what underlies the claim of the Central Dogma—what precludes a reversible flow of information, what makes it unthinkable for such information to pass from proteins to nucleic acid.

Finally, another (closely related) difference between *information* in cyberscience and in molecular biology also needs to be noted. Both in the early paper by Watson and Crick (1953) and later, when it was spelled out so unambiguously in the Central Dogma (Crick 1958), the locus of information was understood to reside exclusively in the DNA, unavailable for distribution beyond the central office.[10] Later, of course, the possibility of genetic

9. This point has recently been emphasized by Sidney Brenner (personal communication, 1993, 1994).

10. Three years earlier, geneticist H. Kalmus had suggested a more explicitly "cybernetic" notion of genetic information, which he embedded in three "cybernetical or integrative systems": genetic, neural, and social. "Nevertheless," he concludes, "no organism, solitary or social, is conceivable, which has not grown under the control of a well-integrated communications system, the elements of which are the genes" (1950:22). And one month earlier, a letter appeared in *Nature*, cosigned by Boris Ephrussi, Urs Leopold, James D.

authority was extended to include RNA as well but never proteins.

Indeed, much the same can be said for Monod and François Jacob, whose operon model of genetic regulation, introduced in 1960, did so much to popularize the notion of feedback among molecular biologists. To be sure, the operon model did grant to nongenetic elements some role in regulation and hence a certain kind of feedback (indeed, Monod and Jacob early on embraced the terminology of both *feedback* and *cybernetics*), but the role they assigned to the regulatory force of proteins was merely quantitative: it allowed only for regulation of the *rate* of protein synthesis; both the program and its means of execution remained confined to the genome itself.[11] As Jacob and Monod wrote in 1961,

> The discovery of regulator and operator genes, and of repressive regulation of the activity of structural genes, reveals that the genome contains not only a series of blue-prints, but a coordinated program of protein synthesis and the means of controlling its execution. (354)

Watson, and Jean J. Weigle (1953:701), suggesting the term *inter-bacterial information* as a way to resolve a terminology problem in bacterial genetics: "It does not imply necessarily the transfer of material substances, and recognizes the possible future importance of cybernetics at the bacterial level."

11. Jean-Paul Gaudilliere (1994) argues persuasively for the importance of distinguishing between Monod and Jacob here, especially given their difference disciplinary practices and histories.

Nineteen sixty-one was a bounty year for molecular biology and for Jacob in particular. With Sidney Brenner and Matthew Meselson, that same year he helped identify the carrier of the messages from the central office; he found the messenger. Everything else—the extranucleic body of the organism—is somatic surplus, designed, constructed, and maintained by the genetic psyche, the all but incidental medium of genetic transmission.

Cyberscience and molecular biology may have been products of the same historical moment, but with respect to their models of causal structure, they were running on two separate tracks, side by side but in opposite directions: while the first one was busy using the organism to illustrate a new kind of machine, the other was seeking to model the organism after the machines of yesteryear. To be sure (as I've already indicated), other movements were also afoot. Systems engineers organized several workshops, conferences, and such to develop the application of information theory to biological systems,[12] assuming (along with Watson and Crick) the primary if not sole locus of biological information was to be the nucleic acid sequence, but they soon abandoned these efforts—in large part out of recognition of the important differences

12. A particularly active figure in these efforts was Henry Quastler. Quastler organized several meetings on the subject beginning in 1949 (see especially Quastler 1953, and Yockey, Platzman, and Quastler 1958).

between *genetic information* and *information* defined as negative entropy.[13]

The distinction Lwoff insists upon is in notable contrast to the conflation that Kalmus had earlier effected when he wrote, "An important similarity between a message and a gene is, that they can both be regarded as 'Gestalten,' i.e., as meaningful structures, and that their loss can be regarded as a loss of 'negative entropy'" (1950:20).

Molecular biologists were particularly quick to write off the utility of such theoretical applications.[14] It might even be argued that they actively resisted such applica-

13. As André Lwoff put it in his Compton lectures in 1960,

The living organism is not only an improbable system but a system fitted for certain functions. Low probability and value are not synonymous. . . .

The biologist feels that the functional order is an essential part of the living system. It is clear, however, that this functional order cannot be measured in terms of entropy units, and is meaningless from a purely thermodynamical point of view. . . . [Thus] the term "information" has for the biologist a different sense than for the physicist. (Lwoff 1965:93)

14. Horace Judson expressed a view widely held by molecular biologists when he wrote, "Information theory made no difference to the course of biological discovery: when the attempt was made to apply it—when a conference was held, a volume published—the mathematical apparatus of the method produced comically little result. The biological papers (and the papers' citations) of the time testify to that by their silence" (1979:244). See also Sahotra Sarkar (1993) and, for a rather different reading, Lily Kay (in press).

15. If my argument about the importance of the difference in meanings of the term *information* in the two disciplines is correct, it might even be argued that their resistance was not only appropriate but necessary. It served to protect that difference and to maintain the full range of ambiguity in their own use of the term.

tions.[15] However, some (distinctly nonmolecular) biologists saw a quite different and perhaps more profitable use to be made of the models the new machine had to offer. Instead of claiming the notion of information to support a highly reductionist and unidirectional causal structure, a small number of biologists interested in the complexities of embryogenesis sought in cybernetics and information theory support for a dynamic and interactive conception of organism. Although these efforts are now largely forgotten, and were not at the time followed up, they are nonetheless worth noting, if only for historical interest.

DEVELOPMENTAL BIOLOGY

One of the first of these was C. H. Waddington, perhaps the leading advocate in his time of the need to integrate the findings of genetics with those of embryology. In an extended discussion of "The Cybernetics of Development," published in 1957 (chapter 2 in *Strategy of the Genes*), he called upon the work of cyberneticians Ross Ashby (1952), K. G. Denbigh, G. Hicks, and F. M. Page (1948) to support an epigenetic account of "canalysed . . . developmental pathways." Regulation—in Waddington's scheme, that which guarantees the stability of developmental pathways to particular end states—depends on the metabolic complexity of "open systems of autocatalytic reactions."[16] Extensive cross-linkages between competing biochemical reactions provide the feedback necessary to keep the system on track. Coining the term *creode* to describe "a pathway of change which is equilibrated in the sense that the system tends to return to it after disturbance," he tacitly invoked the work on self-steering mechanisms out of which cybernetics had grown: "The path followed by a homing missile, which finds its way to a stationary target, is a creode" (Waddington 1957:32). H. Kacser provided further elaboration of the argument in an appendix that began,

> The belief that a living organism is "nothing more" than a collection of substances, albeit a very complex collection of very complex substances, is as widespread as it is difficult to substantiate. . . . What is required is a demonstration that those properties and types of behaviour which we believe to be characteristic of living organisms . . . are "nothing more" than the result of the complex interplay of inanimate substances and processes. The

16. Four years later, in a critique of conventional uses of information theory in biology, Waddington emphasized that although we at present have only a theory applying to the information transfer from gene to cytoplasm, the "traffic is certainly two way" (1961:121). "Perhaps," he concludes, "to use an analogy not much cruder than many that pass muster in information theory, it is only words that can go into solution and become the playthings of biochemists, but sentences remain always within the domain of the morphologist" (126).

problem is therefore the investigation of *systems*. (Waddington 1957:191)

Other attempts in a similar vein followed soon after. In 1961 Christiaan Pieter Raven, professor of zoology in Utrecht and former director of the Hubrecht Laboratory, attempted to reclaim the term *information* for the more global needs of a developmental biologist. He wrote,

The germ at the beginning of development must be provided with the necessary information to do so. In other words: the ordered structure, the pattern of the fertilized egg cell, must be such that it faithfully "represents" the pattern of the organism which develops from the egg. (1961:6)

The principal question he must therefore address is how the developmental information is encoded in the fertilized egg. Relying explicitly on information theory, he sets out to quantify the plausible sources of the information necessary. As did Watson and Crick, he located genotypic information in the structure of DNA, but added two other sites of stored information: the cellular cortex and the cytoplasm. The cytoplasm is of particular importance in view of the need for gene activation to initiate normal development. But from where does this cytoplasmic information come? Raven's answer was vague but to the point: "The whole body of the maternal organism partakes in the encoding of the cytoplasmic information of the egg cell" (199).

One last example: beginning where Waddington left off, Michael Apter, a maverick zoologist with computer expertise, offered a sustained plea for a more thoroughgoing and more rigorous marriage of cybernetics and development, claiming that "developing organisms are exceedingly complex systems, which can only be understood finally in terms of general principles of organisation and control" (1966:33). Indeed, they are so complex that even servomechanisms do not (at least not yet) offer adequate analogues. Accordingly, the problems of growth and development "are not only important to biology: they are important at the more theoretical level to cybernetics itself, or rather to the abstract theory of systems level of cybernetics" (34). Future progress depends on the theoretical development of self-reproducing automata. Rephrasing the fundamental question of embryogenesis, Apter asked,

How can a single automaton, by starting a process of self-reproduction, each new automaton having the possibility of reproducing itself in turn, produce a system much larger than itself of which it is itself only a part, such that this system is organised in terms of a pattern specified by the original automaton? What is required is a design, or rather a set of designs, for a zygote. (50)

Despite their manifest differences (Raven emphasized the twoness of sexual reproduction; Apter elided it), what these authors have in common is their reach to cybernetics, systems dynamics, and automata theory in an attempt to theorize biological development. The bottom line for all three (the message, as it were) was that genes alone cannot suffice to explain the phenomenon of embryogenesis; one way or another, the entire organism, must be involved. Just how, they could not say, but the image of self-steering, self-organizing systems, or automata, seemed to them to offer a compelling model. But not so to others. The simple fact is that by that time, the entire question of embryonic development had lost its appeal for the generation of biologists coming of age (especially in the United States). It was simply too complex. The sixties—the very years in which molecular biology was soaring—were a low point in the history of developmental biology, its lack of palpable success frequently invoked in contrast with the dramatic achievements of molecular biology. Simplicity was the new credo in biology.

Circulation

Still, even while researchers in molecular biology and cyberscience displayed little interest in each other's epistemological program, *information*—either as metaphor or as material (or technological) inscription—could not be contained. In the real world, there was no stopping the circulation of meaning, no cutting of what Lacan calls the circuit of language. In the 1960s, metaphor, not material exchange, provided the primary vehicle for this circulation.In other words, it was the metaphorical use of *information*—as it criss-crossed among these two sets of disciplines, their practitioners, and among their subjects—that provided the principal vector for the dissemination of meaning.

No one in either camp could be unaware of the success and glamour of the other. Perhaps especially, it was not possible for the successes of molecular biology to go unnoticed by the cyberneticians. In 1964 Wiener published *God, Golem, Inc.,* a book intended for a more popular audience than had been his earlier work, *Cybernetics* (1948). In the latter he had argued first for the homology between animals and machines and second for the "mutual convertibility of spatial and functional structure, on the one hand, and of messages in time, on the other"—for the mutual convertibility of animals or machines and messages. But this time around, he started with "the simple fact" we learn from molecular biology:

> In the presence of a suitable nutritive medium of nucleic acids and amino acids, a molecule of a gene . . . can *cause* the medium to lay itself down into other molecules [emphasis added]. . . . It is this act of molecular multiplication . . . which seems to represent a late stage of analysis of the vast and complicated process of reproduction.

Man makes man in his own image. This seems to be the echo or the prototype of the act of creation, by which God is supposed to have made man in His image. Can something similar occur in the less complicated . . . case of the nonliving systems that we call machines? . . . Can the . . . machine itself act as an archetype, even as to its own departures from its own archetypal pattern?

It is the purpose of the present section to answer these questions, and to answer them by 'yes.' . . . Different as the mechanical and biological reproduction may be, they are parallel processes, achieving similar results; and an account of one may well produce relevant suggestions in the study of the other. (29–30)

But what, he asks a few pages later,

is a machine? From one standpoint, we may consider a machine as a prime mover. . . . This is not the standpoint which we shall take in this book. For us, a machine is a device for converting incoming messages into outgoing messages. . . . As the engineer would say in his jargon, a machine is a multiple-input, multiple-output transducer. (31–32)

A few pages further into the text comes the punch line:

The transducer—the machine, as instrument and as message—thus suggests the sort of duality which is so dear to

the physicist. . . . Again, it suggests that biological alter-
nation of generations which is expressed by the *bon mot*
. . . that a hen is merely an egg's way of making another
egg. . . . Thus the machine may generate the message,
and the message may generate another machine.

That is an idea with which I have toyed before—that it
is conceptually possible for a human being to be sent
over a telegraph line. . . . At present, and perhaps for the
whole existence of the human race, the idea is impracti-
cable, but it is not on that account inconceivable. (35–36)

Soon, but not quite yet. In 1964 it may not have been possible to send bodies over a telegraph line, but information
could, and did, flow freely.

Six years later it was evident that Jacob, along with
Lacan—indeed, along with much of the academic world—
had been reading his Norbert Wiener. Reviewing the history
of *The Logic of Life*,[17] Jacob wrote,

This isomorphism of entropy and information establishes
a link between the two forms of power; the power to do
and the power to direct what is done. (1974:251)

But try as he might to contain the modernist subject in the
executive office of the chromosome, he was by then already
behind the times: no such direct link between doing and

17. First published in 1970 in French.

directing was any longer possible. The distributed, dispersed nature of power—both the power to do and the power to direct—had become explicit and grist for new managerial strategies. If power resides in information, how could that information possibly be contained? It would be far better to try exploit its circulation. As Jacob did. And in doing so, he began to waffle. Quoting Wiener extensively, Jacob's account of the mandated one-way flow of genetic information unwittingly began to lose its sense of direction. He wrote:

> With the possibility of carrying out mechanically a series of operations laid down in a programme, the old problem of the relations between animal and machine was posed in new terms. "Both systems are precisely parallel in their analogous attempts to control entropy through feedback," said Wiener. . . . Both have special equipment . . . for collecting at a low energy-level the information coming from the outside world and for transforming it for their own purposes. . . . The coordination of the system depends on a network of regulatory circuits by which the organism is integrated. (1974:252–53)

And finally, with what reads like a total inversion of master and slave, he wrote, "At any time, the machine that executes its programme is capable of directing its action, of correcting or even interrupting, in accordance with the message received" (253).

From Clocks and Telegraphs to Computers

Schroedinger, writing more than twenty-five years earlier, had ended his book with this explicit worry:

> But please do not accuse me of calling the chromosome fibres just the "cogs of the organic machine"—at least not without a reference to the profound physical theories on which the simile is based. . . . The single cog is not of coarse human make, but is the finest masterpiece ever achieved along the lines of the Lord's quantum mechanics. (1944:91)

However, Schroedinger had only clocks and telegraph wires—apart from the Lord's quantum mechanics—to think with. By the 1970s physicists, biologists, and engineers (as well as the rest of us) had a new kind of machine to think with, based not on unidirectional transmission of messages from sender to receiver but on networks and systems: As Wiener put it, "a multiple-input, multiple-output transducer." Can it be any surprise, then, that in the bootstrap process of modeling organisms and machines, each upon the other, not only do organisms and machines come increasingly to resemble each other but that, as they do, the meaning of both terms undergoes some rather critical changes?

As Life went on—with its inevitably circular flow of information and soon of people as well—the polarity (or counterpoint) between cyberscience and molecular biology that had been so sharp in the early years became ever more diffuse. Yet a vital tension remained, only now within as well as between the disciplines. Just as Jacob oscillated between a vision of the DNA as source of all information and direction, and the organism as a network of regulatory circuits, so too did Wiener. His ambivalence (or oscillation) was already evident in *God, Golem, Inc.*. Although explicitly repudiating the conception of "machine as a prime mover," the starting point of his argument was nonetheless the unforgettable lesson from molecular genetics—namely, that "a molecule of a gene . . . can cause the medium to lay itself down into other molecules" (28). Starting with the gene as cause, he moved rapidly to machines as organisms, to self-reproducing automata, and finally to the machine as a message. As Jacob quoted Wiener, the organism can itself be "seen as a message" (Jacob 1974:252).

Of course, for those who cared, there still remained something of a gap between Jacob's notion of message as a molecule and Wiener's notion of a body that can be "sent over a telegraph line." But to see this, one needed to be conversant with both molecular biology and automata theory. In 1969 Howard Pattee had identified the slippage with the simple question, "How Does a Molecule Become a Message?" He wrote,

I am convinced that the problem of the origin of life cannot even be formulated without a better understanding of
how molecules can function symbolically, i.e., as records,
codes, and signals. . . . We need to know how a molecule
becomes a message. . . . Nothing I have learned from molecular biology and genetics tells me in terms of basic
physical principles why matter should ever come alive. (1)

Turning to the already extensive studies of artificial language, he concluded that

a molecule does not become a message because of any
particular shape or structure. . . . A molecule becomes a
message only in the context of a larger system of physical constraints which I have called a "language." (8)

And his closing prognosis:

To understand how the molecules became messages . . .
will require a much deeper appreciation of the physics
of switches and the logic of networks. (15)

Pattee's question of how a molecule becomes a message
has not yet been answered, but something else of note has
begun to happen. His remarks were published in the supplement to a new journal that had been launched just three
years earlier to mark the first stirrings of a rebirth of the classical conundrum of biology—namely, the problem of
embryogenesis, albeit under the manifestly more neutral

term of *developmental biology.* Other journals and conference proceedings followed. The first Gordon Conference on Developmental Biology was held in 1972; the same year, *Peterson's Annual Guide to Graduate Programs in the Biological Sciences* listed its first program in developmental biology; Nüsslein-Volhard published the first of her papers on maternal effects in embryological development in 1977; Ed Lewis's classic paper synthesizing thirty years of work on the complex of genes associated with the organization of body parts appeared in *Drosophila* in 1978. By 1980 *Peterson's Guide* listed fifty programs in developmental biology. Over the next ten years, developmental biology became one of the, if not the, prime research areas of the biological sciences.

The effect of this shift in research priorities has been to radically subvert the dogma of gene action (see chapter 1). To quote Francisco Varela and Jean-Pierre Dupuy, molecular biology had seemed to be

a model success in the reduction of life to macromolecular chemistry, mainly through the discovery of the genetic code and the notion of a cellular programming which is supposed to stand at the base of all development as it (literally) writes the organism as it unfolds in its ontogeny. However, after an initial phase of enchantment with the idea, it has become clear . . . that if one takes the notion of a genetic program literally one falls into a strange loop:

one has a program that needs its own product in order to be executed. In fact, every step of DNA maintenance and transcription is mediated by proteins, which are precisely what is encoded. To carry on the program it must already have been executed! (1992:4)

Varela and Dupuy's point is to suggest this problem of development—the circular logic of chicken and egg—as paradigmatic of "the logic of the supplement." But they might just as easily have invoked the circular logic of a cybernetic paradigm.

The Medium as Message

Driven by the exigencies of postwar political and institutional dynamics on the one hand and technical developments in both molecular biology and cyberscience on the other, the circulating economy of machines, messages, and organisms provided new ways of talking about the emergence of biological form hand in hand with new ways of talking about the forms and management of social and technological structures. In particular, it provided a language for describing the distribution of power in messy, and conspicuously heterogeneous, organisms, of thinking about the functional circuitry of systems in which the elementary constitutive units (the cells) are clearly not identical, even if their DNA is.

In the late seventies, a more strictly internal technical

development brought about its own transformation in the conventional programs of molecular biology. Techniques of recombinant DNA created a new order of opportunities for cloning and sequencing, for propelling molecular biology ever further toward becoming an informational science, a science of command, control, and communication. But the real decade of turnover—in molecular biology, in cyber-science, in popular culture, and even in physics—was the eighties. It was in the eighties that the winds of information became a whirlwind, and the classical body—of embryology, even of the material universe—lost its mooring.

A few benchmarks from the larger culture are worth noting: the term *cyberspace* was coined in 1984 (Gibson); 1984 was also the year of Danny Hillis's "Connection Machine" and the launching of a Cambridge, Massachusetts-based coalition of computer scientists and physicists called "Thinking Machines," incorporated to build parallel computers. And It was also the year that physicist John Wheeler published his first paper on the universe as a computer. But whether physicists yet saw the computer as a persuasive model for the universe, among biologists it became widely accepted as a suitable model for the cell (or organism). Consider the following description of the process of embryogenesis in *C. elegans,* as it was presented in one of the most widely used textbooks of the late eighties, *The Molecular Biology of the Cell:*

For cells, as for computers, memory makes complex programs possible; and many cells together, each one step-
ping through its complex developmental control program, generate a complex adult body. . . . Thus the cells
of the embryo can be likened to an array of computers
operating in parallel and exchanging information with
one another. (Alberts et al. 1989:902)

Molecular biologists of Delbrück's generation are probably still not much interested in information processing,
but of necessity a younger generation is. The sheer prolif-
eration of data requires it. Indeed, partly in response to the
call that Eric Lander and Charles Cantor issued to computer scientists in 1988 on behalf of the Human Genome
Project, a new field has begun to emerge that recognizes
no boundary at all between information processing and
molecular biology. True, molecular biology is still a wet
science, but techniques that permit the rapid cloning of an
arbitrary piece of DNA, given only the sequences of the
end portions (polymerase chain reaction, or PCR), cou-
pled with the transmission of sequence data by electronic
mail, are clearly moving it into a new domain. As Walter
Gilbert puts it, biology is moving "toward a paradigm
shift" (1991): in Gilbert's vision of the new paradigm, the
reagent of the future will be sequence data. However,
those with more sophistication in cyberscience see the
fundamental building blocks as networks. If Pattee's

analysis is correct, the molecule is well on its way to becoming a message.

The term *beaming* is of course a science fiction term (introduced to popular culture by the first generation of "Star Trek"). But as we well know, it is in scientific literature that most science fiction originates, and indeed we saw the idea already being proposed explicitly by Norbert Wiener as early as 1964. In arguing for interchangeability between organisms and the particular kind of machines known as multiple-input, multiple-output transducers, Wiener was not simply erasing the distinction between bodies and machines; he was also relying on the conflation between machines and messages that has been endemic in computer discourse since its very beginnings. But the late eighties brought at least four new dimensions to the vision of sending human beings over the wire, and all appeared about the same time. The most familiar are virtual reality and artificial life (see Doyle 1993), and I would add two new techniques from molecular biology: nucleotide sequencing and polymerase chain reaction (PCR).[18]

18. I might add one last development, perhaps IBM's hottest contribution yet to the circuit of infomatic exchange, to this sequence. The April 15, 1993, issue of *Nature* reports:

The latest idea from science fiction to be adopted by quantum physicists . . . is teleportation. . . . The idea behind teleportation is that a physical object is equivalent to the information needed to construct it; the object can therefore be transported by transmitting the information along any conventional channel of telecommunication, the receiver using the information to reconstruct the object. (Sudbury 1993:586)

As usual, "Star Trek" is not far behind (nor, for that matter, is it far ahead): in a particularly brilliant recent episode, a prototype of Sherlock Holmes's Professor Moriarty, recreated on the *Enterprise*'s Holodeck, acquires consciousness. And with the acquisition of consciousness, he demands—

But unlike a gene, a single quantum cannot be cloned. It may be for this reason that the accompanying cartoon draws not so much on electronically mediated cloning as it does on a more traditional genre of experiments in developmental biology focused on the organization (or reorganization) of body parts.

in fact, asserts—his existence. Overriding his creators' demurrals about the difference between simulated and real existence, he leaps out of cyberspace, declaring, of course, "Cogito ergo sum!" But the self behind Moriarty's cogito is clearly not his own, certainly no more his own than the self behind Descartes' cogito can any longer be said to have been his own; that, too, is a simulacrum: a message from somewhere, from where, or from whom, is no longer the question. Perhaps we no longer even care. Finally, can we even tell whether Moriarty has come alive in virtual or in three-dimensional (as we still say, real) space? And does even this difference still matter? Or perhaps we should rather ask, to whom, and for what, has it stopped mattering?

❋

Certainly, in the life sciences, it matters less and less. Owing in large part to the reemergence of developmental biology, molecular biology may well be said to have "discovered the organism," but the subject of the new biology, however whole and embodied, is only a distant relative to the organism that had occupied an earlier generation of embryologists. As a consequence of the technological and conceptual transformations we have witnessed in the last three decades, the body itself has been irrevocably transformed, perhaps especially in biological discourse. Today's biological organism bears little resemblance to the traditionally maternal guarantor of vital integrity, the source of nurture and sustenance; it is no longer even the passive material substrata of classical genetics. The body of modern biology, like the DNA molecule—and also like the modern corporate or political body—has become just another part of an informational network, now machine, now message, always ready for exchange, each for the other.

As Lacan so rightly observed, "It is very odd to say, . . . man *has* a body" (1988:72). But as he also observed, this locution has been around for a long time—probably since Descartes's 1664 essay "On Man." "Flip through it," Lacan suggests, "and confirm that what Descartes is looking for in man is the clock" (74).

Well, we've come a long way from the wind-up clock, even from the great nineteenth-century developments of the electric motor, the steam engine, and the telegraph. Electricity has given way to electronics and matter and energy to information. In the late twentieth century, it is the computer that dominates our imagination, and it has liberated us from that odd locution "man has a body." In its place we have an even odder set of locutions. Today, it might be more correct to say that the body—in the sense that the word has now acquired—has man. And this body may well have man in a grip tighter than any maternal body ever did.

Alberts, Bruce et al. 1989. *Molecular Biology of the Cell*. 2d ed. New York: Garland.

Allen, Garland 1986. "T. H. Morgan and the Split Between Embryology and Genetics, 1910–1926." In T. J. Horder, I. A. Witkowski, and C. C. Wylie, eds., pp. 113–46, *A History of Embryology*. Cambridge, England: Cambridge University Press.

Apter, Michael. 1966. *Cybernetics and Development*. Oxford, England: Pergamon.

Ashburner, Michael. 1990. "Puffs, Genes, and Hormones Revisited." *Cell* 61 (1): 1–3.

Ashby, Ross. 1952. *Design for a Brain*. London: Chapman and Hall.

Atlan, Henri and Moshe Koppel. 1990. "The Cellular Computer DNA: Program or Data." *Bulletin of Mathematical Biology* 52 (3): 335–48.

Austin, J. L. 1962. *How to Do Things with Words*. Cambridge, Mass.: Harvard University Press.

Baltimore, David. 1984. "The Brain of a Cell." *Science 84* (November): 149–51.

Baltzer, Fritz. 1967. *Theodor Boveri*. Trans. by Dorthea Rudnick. Berkeley: University of California Press.

Beardsley, Tim. 1991. "Smart Genes." *Scientific American* (August): 87–95.

Beer, Gillian. 1989. "'The Death of the Sun': Victorian Solar Physics and Solar Myth." In J. B. Bullen, ed., pp. 159–80, *The Sun Is God: Painting, Literature, and Mythology in the Nineteenth Century*. Oxford, England: Clarendon.

Blau, Helen M. and David Baltimore. 1991. "Differentiation Requires Continuous Regulation." *Journal of Cell Biology* 112 (5): 781–83.

Boycott, A. E. et al. 1930. "The Inheritance of Sinistrality in *Limnaea peregra*." *Philos. Trans. of the Royal Society*, Series B, 219:51–130.

Brink, R. A. 1927. "Genetics and the Problems of Development." *American Naturalist* 61 (574): 280–83.

Campbell, Lewis and William Garnett. 1882. *The Life of James Clerk Maxwell*. London: Macmillan.

Crick, Francis. 1958. "On Protein Synthesis." *Symposium of the Society of Experimental Biology* 12:138–63.

___. 1966. *Of Molecules and Men*. Seattle: University of Washington Press.

Denbigh, K. G., M. Hicks, and F. M. Page. 1948. "The Kinetics of Open Reaction Systems." *Transactions of the Faraday Society* 44:479–94.

Dolbear, A. E. 1894. "Life from a Physical Standddpoint." Woods Hole Biological Lectures, Marine Biological Laboratories, Woodshole, Mass., pp. 1–21.

Donnan, F. G. 1934. "Activities of Life and the Second Law of

Thermodynamics." Letters to the Editor, *Nature* (January 20): 99. (See also Jeans, Sir James H.)

Donnan, F. G. and E. A. Guggenheim. 1934. "Activities of Life and the Second Law of Thermodynamics." Letters to the Editor, *Nature* (April 7): 530; (June 9): 869; (August 18): 255. (See also Jeans, Sir James H.)

Doyle, Richard. In press. *On Beyond Living: Rhetorics of Vitality and Post Vitality in Molecular Biology.* Palo Alto, Calif.: Stanford University Press.

Dunn, L. C. 1959. Transcript of oral history memoir. Oral History Office, Butler Library, Columbia University.

Eddington, Sir Arthur Stanley. 1928. *The Nature of the Physical World.* Cambridge, England: Cambridge University Press.

Edwards, Paul. In press. *The Closed World: Computers and the Politics of Discourse in Cold War America.* Cambridge, Mass.: MIT Press.

Ephrussi, B. et al. 1953. "Terminology in Bacterial Genetics." *Nature* 4355 (April 18): 701.

Galison, Peter. In press. "The Ontology of the Enemy." *Critical Inquiry.*

Garcia-Bellido, A. and J. B. Merriam. 1969. "Cell Lineage of the Imaginal Discs in *Drosophila* Gynandromorphs." *Journal of Experimental Zoology* 170 (1): 61–75.

Gaudilliere, Jean-Paul. 1994. "Information or Regulation: The Rhetoric and Practice of Molecular Biology in and out of the Pasteur Institute." Paper presented at the 1994 Summer Academy of the Verbund fur Wissenschaftsgeschichte, Berlin.

Gibson, William. 1984. *Neuromancer.* New York: Ace Books.

Gilbert, Walter. 1991. "Towards a Paradigm Shift in Biology." *Nature* 349:99.

Goldschmidt, Richard. 1932. "Genetics and Development." *Biological Bulletin* 63 (3): 337–56.

———. 1940. Letter to L. C. Dunn, May 27. L. C. Dunn Papers, American Philosophical Library, Philadelphia.

Hacking, Ian. 1982. "Language, Truth, and Reason." In M. Hollis and S. Lukes, eds., pp. 48–66, *Rationality and Relativism*. Cambridge, Mass.: MIT Press.

Haecker, V. 1926. "Phänogenetisch gerichtete Bestrebungen in Amerika" [Phenogenetic Directed Efforts in America]. *Z. indukt. Abst. Vererb.* 41:232–38.

Haraway, Donna. 1981–82. "The High Cost of Information in Post–World War II Evolutionary Biology: Ergonomics, Semiotics, and the Sociobiology of Communication Systems." *Philosophical Forum* 13 (2–3): 244–78.

Harrison, Ross. 1937. "Embryology and Its Relations." *Science* 85:369–74.

Hartley, R. V. L. 1928. "Transmission of Information." *Bell System Technical Journal* 7:526–63.

Harwood, Jonathan. 1993. *Styles of Scientific Thought: The German Genetics Community, 1900–1933*. Chicago: University of Chicago Press.

Heims, Steve J. 1973. Letter from Max Delbrück, September 18. Photostat sent to the author in 1992.

———. 1991. *The Cybernetics Group*. Cambridge, Mass.: MIT Press.

Hesse, Mary. 1988. "Theories, Family Resemblances and Analogy." In D. H. Helman, ed., pp. 317–40, *Analogical Reasoning*. Dordrecht, The Netherlands: Kluwer.

Horder, T. J. and P. J. Weindling. 1986. "Hans Spemann and the Organiser." In T. J. Horder, J. A. Witkowski, and C. C. Wylie, eds., pp. 183–242, *A History of Embryology*. Cambridge, England: Cambridge University Press.

Hughes, Thomas P. 1993. "Modern and Postmodern Engineering." Paper presented at Seventh Annual Arthur Miller Lecture on Science and Ethics, April 8, Massachusetts Institute of Technology, Cambridge, Mass.

International Congress of Genetics. 1941. *Proceedings* of the seventh annual congress, 1939, Edinburgh. Cambridge, England: Cambridge University Press.

Jacob, François. 1974. *The Logic of Life*. New York: Pantheon Press.

Jacob, François and Jacques Monod. 1961. "Genetic Regulatory Mechanisms in the Synthesis of Proteins." *Journal of Molecular Biology* 3:318–56.

Jeans, Sir James H. 1934. "Activities of Life and the Second Law of Thermodynamics." Letters to the Editor, *Nature* (February 3): 174; (April 21): 612; (June 30): 986. (Interaction with F. G. Donnan.)

Johnstone, James. 1914. *The Philosophy of Biology*. Cambridge, England: Cambridge University Press.

Judson, Horace. 1979. *The Eighth Day of Creation*. New York: Simon & Schuster.

Kalmus, H. 1950. "A Cybernetic Aspect of Genetics." *Journal of Heredity* 41 (1): 19–22.

Kay, Lily. Forthcoming. "Who Wrote the Book of Life?" In M. Hagner and H. J. Rheinberger, eds., *Experimentisystem in den Biologishe-Medizinishcen Wissenschaften* [Experimental Systems in the Bio-Medical Sciences]. Berlin: Akademie Verlag.

Keller, Evelyn Fox. 1990. "Physics and the Emergence of Molecular Biology." *Journal of the History of Biology* 23 (3): 389–409.

___. 1994. "Rethinking the Meaning of Genetic Determinism." In *Tanner Lectures on Human Values*. Salt Lake City: University of Utah Press.

Knott, C. G. 1911. *Life and Scientific Work of Peter Guthrie Tait*. Cambridge, England: Cambridge University Press.

Lacan, Jacques. 1988. *The Seminars of Jacques Lacan*, edited by J. A. Miller. New York: Norton.

Leff, Harvey S. and Andrew F. Rex. 1990. *Maxwell's Demon*. Princeton, N.J.: Princeton University Press.

Lewin, Roger. 1984. "Why Is Development So Illogical?" *Science* 224:1327–29.

Lewis, G. N. 1930. "The Symmetry of Time in Physics." *Science* 71:569–77.

Lewontin, Richard. 1992. "The Dream of the Human Genome." *New York Review of Books*, May 28, pp. 31–40.

Lillie, Ralph. 1914. "The Philosophy of Biology: Vitalism Versus Mechanism." *Science* 60 (1041): 840–46.

Lwoff, André. 1965. *Biological Order*. Cambridge, Mass.: MIT Press.

Martin, Emily. 1991. "The Egg and the Sperm: How Science Has Constructed a Romance Based on Stereotypical Male-Female Roles." *Signs: Journal of Women in Culture and Society* 16 (3): 485–501.

Medawar, Sir Peter. 1965. "A Biological Retrospect." *Nature* 207 (5004): 1327–30.

Morgan, T. H. 1924. "Mendelian Heredity in Relation to Cytology." In E. V. Cowdry, ed., pp. 691–734, *General Cytology*. Chicago: University of Chicago Press.

___. 1926a. *The Theory of the Gene*. New Haven: Yale University Press.

___. 1926b."Genetics and the Physiology of Development." *American Naturalist* 60 (671): 489–515.

___. 1934. *Embryology and Genetics*. New York: Columbia University Press.

Muller, H. J. 1929. "The Gene as the Basis of Life." Paper presented at symposium entitled The Gene. International Congress of Plant Sciences, Section of Genetics, August 19, 1926, Ithaca, N.Y. Published in *Proceedings of the International Congress of Plant Science, I*. Menasha, Wisc.: George Banta Publishing, pp. 897–921.

Needham, Joseph. 1943. *Time: The Refreshing River*. London: Allen & Unwin.

Nijhout, H. F. 1990. "Metaphors and the Role of Genes in Development." *Bioessays* 12 (9): 441–46.

Pattee, Howard. 1969. "How Does a Molecule Become a Message?" *Developmental Biology Supplement* 3:1–16.

"Progress Report of the Air Defense Systems Engineering Committee." May 1, 1950. Written for the science advisory board to the chief of staff, U.S. Air Force, "George Vallee Committee," C50–10788-AF.

Quastler, Henry, ed. 1953. *Essays on the Use of Information Theory in Biology*. Urbana: University of Illinois Press.

Raven, Christiaan Pieter. 1961. *Oogenesis: The Storage of Developmental Information*. Oxford, England: Pergamon.

Rosch, Eleanor and C. B. Mervis. 1975. "Family Resemblances:

Studies in the Internal Structure of Categories." *Cognitive Psychology* 7:573–605.

Rosenblueth, Arturo, Norbert Wiener, and Julian Bigelow. 1943. "Behavior, Purpose, and Teleology." *Philosophy of Science* 10:18–24.

Sager, Ruth and Francis Ryan. 1961. *Cell Heredity*. New York: Wiley.

Sander, Klaus. 1986. "The Role of Genes in Ontogenesis—Evolving Concepts from 1883 to 1983 as Perceived by an Insect Embryologist." In T. J. Horder, J. A. Witkowski, and C. C. Wylie, eds., pp. 363–95, *History of Embryology*. Cambridge, England: Cambridge University Press, 1986.

Sapp, Jan. 1987. *Beyond the Gene*. Oxford, England: Oxford University Press.

Sarkar, Sahotra. 1993. "Biological Information: A Skeptical Look at Some Central Dogmas of Molecular Biology." Paper presented at Thirty-Third Annual Boston Colloquium for the Philosophy of Science, April 13, Boston University.

Schroedinger, Erwin. 1935–43. Letters to F. G. Donnan, August 25, 1935; September 7, 1939; October 26, 1942; December 15, 1942; January 24, 1943; September 21, 1943; October 24, 1943; December 23, 1943; January 5, 1946; and March 25, 1946. Manuscripts and Rare Books, University College Library, London.

___. 1944. *What Is Life?* Cambridge, England: Cambridge University Press.

___. 1992. "Autobiographical Sketches." In *What Is Life? The Physical Aspect of the Living Cell*, November 1960. Canto ed., Cambridge, England: Cambridge University Press.

Schweber, Silvan S. 1982. "Angels, Demons, and Probability: Some Aspects of British Science in the Nineteenth Century." In A. Shimony and H. Feshbach, eds., pp. 319–63, *Physics as Natural Philosophy: Essays in Honor of Laszlo Tisza on His Seventy-Fifth Birthday.* Cambridge, Mass.: MIT Press.

Shannon, C. E. 1948. "A Mathematical Theory of Communication." *Bell System Technical Journal* 27:379–479, 623–56.

Smith, Crosby and Norton Wise. 1989. *Energy and Empire: A Biographical Study of Lord Kelvin.* Cambridge, England: Cambridge University Press.

Smith, John Maynard. 1989. "Weismann and Modern Biology." In P. H. Harvey and L. Partridge, eds., pp. 1–12, *Oxford Surveys in Evolutionary Biology*, vol. 6. Oxford, England: Oxford University Press.

Spemann, Hans. 1924. "Vererbung und Entwicklungsmechanik" [Heredity and Developmental Mechanics]. *Z. Indukt. Abstammungs-und Vererbungslehre* 33:272–94.

Stewart, Balfour. 1873. *The Conservation of Energy, Being an Elementary Treatise on Energy and Its Laws.* London: H. S. King.

Sturtevant, Alfred H. 1932. "The Use of Mosaics in the Study of the Developmental Effects of Genes." *Proceedings of the Sixth International Congress of Genetics.* New York: Macmillan, p. 304.

Sturtevant, Alfred H. and G. W. Beadle. 1962. *An Introduction to Genetics*, 1939. Reprint, New York: Dover.

Sudbery, Tony. 1993. "Instant Teleportation." *Nature* (April 15): 586–87.

Thomson, William, Baron Kelvin. 1911. *Mathematical and Physical Papers.* Cambridge, England: Cambridge University Press.

Timoféeff-Ressovsky, N. W., K. C. Zimmer, and Max Delbrück. 1935. "Über die Natur der Genmutation und der Genstruktur" [On the Nature of Genetic Mutation and Gene Structure]. *Nachr. Ges. Wiss. Gottingen, Math.-Phys. Klasse*:189–245.

Varela, Francisco, and Jean-Pierre Dupuy, eds. 1992. *Understanding Origins: Contemporary Views on the Origin of Life, Mind, and Society*. Boston: Kluwer.

Waddington, C. H. 1957. *Strategy of the Genes*. New York: Macmillan, 1957.

___. 1961. "Architecture and Information in Cellular Differentiation." In J. A. Ramsay and V. B. Wigglesworth, eds., pp. 117–26, *The Cell and the Organism*. Cambridge, England: Cambridge University Press.

Watson, David L. 1931. "Biological Organization." *Quarterly Review of Biology* 6:143–56.

Watson, J. D. and F. Crick. 1953. "Genetical Implications of the Structure of Deoxyribonucleic Acid." *Nature* 171:964–67.

Weaver, Warren. 1949. "Problems of Organized Complexity." *American Scientist* 36:536–44.

Wiener, Norbert. 1948. *Cybernetics: or, Control and Communication in the Animal and Machine*. Cambridge, Mass.: MIT Press.

___. 1964. *God, Golem, Inc.* Cambridge, Mass.: MIT Press.

Wilson, E. B. 1896. *The Cell in Development and Heredity*. New York: MacMillan, 1896.

___. 1932. "Opening Address." *Proceedings of the Sixth International Congress of Genetics*. New York: Macmillan, p. 82.

Yockey, Hubert, Robert L. Platzman, and Henry Quastler, eds. 1958. *Symposium on Information Theory in Biology*. Oxford, England: Pergamon.

Yoxen, E. J. 1979. "Where Does Schroedinger's 'What Is Life?' Belong in the History of Molecular Biology?" *History of Science* 62:17–52.

Zuckerman, Harriet and Joshua Lederberg. 1986. "Forty Years of Genetic Recombination in Bacteria: Postmature Scientific Discovery?" *Nature* 324 (6098): 629–31.

INDEX

Designer: Teresa Bonner

Text: Optima

Compositor: Columbia University Press

Printer: Edwards Brothers

Binder: Edwards Brothers